Killing Bugs for Business and Beauty

Canada's Aerial War against Forest Pests, 1913–1930

Mark Kuhlberg

Killing Bugs for Business and Beauty examines the beginning of Canada's aerial war against forest insects and how a tiny handful of officials came to lead the world with a made-in-Canada solution to the problem.

Shedding light on a largely forgotten chapter in Canadian environmental history, Mark Kuhlberg explores the theme of nature and its agency. The book highlights the shared impulses that often drove both the harvesters and the preservers of trees, and the acute dangers inherent in allowing emotional appeals instead of logic to drive environmental policy-making. It addresses both inter-governmental and intra-governmental relations, as well as pressure politics and lobbying. Including fascinating tales from Cape Breton Island, Muskoka, and Stanley Park, *Killing Bugs for Business and Beauty* clearly demonstrates how class, region, and commercial interests intersected to determine the location and timing of aerial bombings.

At the core of this book about killing bugs is a story, infused with innovation and heroism, of the various conflicts that complicate how we worship wilderness.

MARK KUHLBERG is a professor and MA Coordinator in the Department of History at Laurentian University and is a leading authority on Canada's forest history.

KILLING BUGS *for* BUSINESS *and* BEAUTY

Canada's Aerial War against
Forest Pests, 1913–1930

Mark Kuhlberg

UNIVERSITY OF TORONTO PRESS
Toronto Buffalo London

ISBN   978-1-4875-0897-5 (cloth)   ISBN   978-1-4875-3943-6 (EPUB)
ISBN   978-1-4875-2647-4 (paper)   ISBN   978-1-4875-3942-9 (PDF)

Library and Archives Canada Cataloguing in Publication

Title: Killing bugs for business and beauty : Canada's aerial war against forest pests, 1913–1930 /
Mark Kuhlberg.
Names: Kuhlberg, Mark, 1966– author.
Description: Includes bibliographical references and index.
Identifiers: Canadiana (print) 20210334630 | Canadiana (ebook) 2021033469X |
ISBN 9781487508975 (cloth) | ISBN 9781487526474 (paper) | ISBN 9781487539436 (EPUB) |
ISBN 9781487539429 (PDF)
Subjects: LCSH: Aerial spraying and dusting in forestry – Canada – History – 20th century. | LCSH:
Insect pests – Control – Canada – History – 20th century. | LCSH: Environmental policy –
Canada – History – 20th century.
Classification: LCC SB764.C3 K84 2021 | DDC 634.9/69097109041–dc23

We wish to acknowledge the land on which the University of Toronto Press operates. This land is
the traditional territory of the Wendat, the Anishnaabeg, the Haudenosaunee, the Métis, and the
Mississaugas of the Credit First Nation.

University of Toronto Press acknowledges the financial support of the Government of Canada,
the Canada Council for the Arts, and the Ontario Arts Council, an agency of the Government of
Ontario, for its publishing activities.

Canada Council
for the Arts

Conseil des Arts
du Canada

ONTARIO ARTS COUNCIL
CONSEIL DES ARTS DE L'ONTARIO
an Ontario government agency
un organisme du gouvernement de l'Ontario

Funded by the   Financé par le
Government   gouvernement
of Canada   du Canada   Canada

To Carling and Nolan, for all the joy they have brought to my life

Contents

Images

Maps

Images

Acknowledgments

I will always remember the first time I learned about bugs … real bugs. Although I had endured small swarms of various critters – mostly pesky mosquitoes – while spending time at the cottages of friends and relatives, nothing had prepared me for the onslaught that I would face when I began treeplanting in northern Ontario. A bad bug day in the cutovers around Dryden when I was an eighteen-year-old rookie meant hiding under a silvicool tarp (intended to shield the seedlings from the hot sun) just so I could gobble down my lunch. To do otherwise would have resulted in eating more black flies than sandwich. And surviving these pests during the time spent planting presented one with a daily decision that involved choosing between two equally uninviting options: lather one's exposed skin in the infamous Muskol repellent, which had a wretched stench and tended to burn holes in the trucks' upholstery, or wrap as much skin as possible in any kind of cloth, which left one sweltering. And of course, to the treeplanters' eternal dismay, once the black flies had finally started to die back, out came the deer and horse flies, whose bites tore chunks from one's flesh and against whom the aforementioned defences were useless.

When treeplanters east of the Rockies talk about the how the biggest challenge they confront in their work is mental, these grizzled bushpeople are talking not about the stick-to-itiveness that comes with counting their daily incomes in eight-and-a-half cent increments or overcoming all types of weather and terrain to make their money. No, they are referring to the bugs, which have driven more than a few planters right out of the bush – and out of their minds – and onto the sanctity of the Greyhound (when it still used to run across our country) that could return them to the safety of city life. Maybe that is why I enjoy winter so much in northern Ontario: no bugs.

All that to say that it was a pleasure to work on this book about forest pests from the screened-in environments of my home office and various archives, and a good many other forces contributed to the enjoyment I derived from researching and writing it. I was very fortunate to receive a Standard Research Grant back in 2010 (410–2010–1661) from the Social Sciences and Humanities Research Council of Canada to support this project. This award permitted me to travel to institutions across North America to investigate the story that I tell in this book. Along the way a long list of archivists, helpful persons, and librarians assisted me in my journey and greatly facilitated it. They work at or are associated with the following institutions or organizations: Abitibi-Bowater Archives (Iroquois Falls, Ontario); Ahmic Lake Cottagers' Association (Magnetawan, Ontario); Archives of Ontario (Toronto); BC Forest Service, BC Legislative Library, and BC Ministry of the Environment (Victoria, BC); Beaton Institute (Sydney, Nova Scotia); Bibliothèque et Archives nationales du Québec (Quebec City); California Academy of Sciences (San Francisco, California); City of Vancouver Archives; Collier County Museums (Naples, Florida); Cornell University – Division of Rare and Manuscript Collections (Ithaca, New York); Directorate of History and Heritage – Canadian Department of National Defence (Ottawa); Entomological Society of BC; F. Franklin Moon Library – College of Environmental Science and Forestry, State University of New York (Syracuse, New York); Forest History Society Archives (Durham, North Carolina); Gelman Library (Special Collections Research Center), George Washington University (Washington, DC); Harpers Ferry Center for Media Services – United States National Parks Service (Harpers Ferry, West Virginia); Iowa State University – Special Collections and Archives (Ames, Iowa); Library and Archives Canada (Ottawa); Maine Historical Society (Portland, Maine); Muskoka Lakes Museum (Port Carling, Ontario); National Film Board (Montreal); National Geographic; Natural Resources Canada – Pacific Forestry Centre (Victoria, BC); Nova Scotia Archives and Nova Scotia Department of Natural Resources Library (Halifax, Nova Scotia); Royal BC Museum and Archives (Victoria, BC); Smithsonian Institute – Department of Entomology (Washington, DC); State Historical Society of Missouri (Columbia, Missouri); United States Forest Service – Region 5 (Vallejo, California); United States National Archives and Records Administration (Washington, DC, and Kansas City, Missouri); University of Georgia – College of Agricultural and Environmental Sciences (Athens, Georgia); University of Idaho Library (Moscow, Idaho); University of Minnesota Libraries (Minneapolis, Minnesota); University of Toronto – Archives and Fisher Rare Book Library; University of Washington (Seattle, Washington); University of Waterloo Archives (Waterloo, Ontario); University of Winnipeg Archives; Washington Forest Protection Association (Olympia, Washington); Washington University in St. Louis (St. Louis, Missouri); West Virginia Historical Archives & Manuscript Collections (Morgantown, West Virginia); Wisconsin Department of Natural Resources (Madison, Wisconsin); Wisconsin Historical Society Archives (Madison, Wisconsin); and Yale University Library – Manuscripts and Archives (New Haven, Connecticut).

I wish to acknowledge formally a number of the individuals who contributed to this book. At the top of the list are a handful of people at the Natural Resources Canada – Pacific Forestry Centre in Victoria, BC. This special shout-out goes to Linda Brown, Mary Kossak, Jeannette Lum, Brad Stennes, and Colin Wood, who were able to locate both documents and photographs that were invaluable to understanding the aerial dusting efforts in British Columbia; Jeannette went above and beyond the call of duty in helping me obtain better copies of images years after she had retired. At Laurentian University, Lina Beaulieu and her inter-library loan office procured many obscure books and reports, Léo Larivière worked his magic to produce the maps that were needed for this effort, and the Laurentian University Research Fund supported the publication of this book. A number of research assistants also pitched in to lend a hand at various times including Julian Colilli, Michael Commito, Rick Duthie, Michael Fantini, Nick McMullen, Scott Miller, Jackson Pind, and Dan Walter. Larissa Kelly was invaluable at the California Academy of Sciences Archives, where she painstakingly took photos of Ralph Hopping's papers and a few other germane documents; she was a pleasure to work with. Teresa Boyd did similar work at the Smithsonian Institute in Washington, DC, and Dr. Garth Reese and his staff were fantastic in facilitating my work at the Special Collections and Archives at the University of Idaho Library. The same is true of Michele Christian, collections archivist at Iowa State University, and Doug Smith at the Muskoka Lakes Museum. The superb crew of archivists, librarians, and historians at the Forest History Society in Durham, North Carolina, also assisted my work on this project. Officials at the office of Senator Susan Collins of Maine were gracious enough to respond to my requests and attempt to answer them, and I will never forget the help the late Marc Dube, mill manager at the old St. Mary's Paper Company in Sault Ste. Marie, Ontario, gave me many years ago. At the University of Toronto Press, I would like to thank Len Husband for his support, the editorial team of Beth McAuley and Samantha Rohrig at The Editing Company, and also the two anonymous reviewers – particularly Reader A-2 – who provided immensely helpful feedback on an earlier draft of this manuscript.

A slew of others helped in ways that were far more personal than professional. My long-time friends in southern Ontario have stood by me every step of the way over the last forty-odd years. They frequently checked in on and encouraged me when the slogging in academia was tough, and did all they could to sustain my spirits; I will never forget them for standing by me all these years. In this respect, Steve Cooke, Sean Sleeth (thanks for proofreading yet another manuscript, SP), Lisa Douglas, and others have been the best friends for whom someone could have asked; it has also been a privilege to get to know Paul Mark over the last decade or so. Chris O'Brien has been a chum for decades, and a few years back reminded what a great person he is by going above and beyond in entertaining my son, Nolan, and his buddy, Cal, when we were based in southern Ontario for a week. Mike Sherwin entered my world when I still carried a comb, and he quickly became a wonderful friend. He continues to share the wonders of his world with my family and me, and for that we are so grateful. Likewise, Ken Armson, now rightfully invested with the Order of Canada, has been a constant source of inspiration and support ever since I first contacted him roughly three decades ago. He has taught me a great deal about the woods – and life. Ken founded the Forest History Society of Ontario over a decade ago, and it is still going strong. It is dedicated to preserving and promoting our province's forest history, and I benefit from my association with it; I have learned so much from the wonderful people who support this cause. I have also erred in failing to acknowledge in previous books my dear Aunt Ilme and Uncle Hans Sepp. They were two folks who injected – seemingly as often as they could – rich doses of happiness into the challenging times that punctuated my childhood. Their acts spanned the gamut of remembering me on their many trips across the United States (whose aunt walks miles to find a postcard of Fenway Park in Boston even though she cares little about baseball?) to making my time up in Georgian Bay that much more enjoyable. I will never forget that they taught me the importance of working in the morning and fishing after lunch. Aitäh tädi Ilme ja onu Hans. Aitäh.

My family and I have called Sudbury home now for nearly twenty years, and the area has provided us with a bounty of wonderful friends who treat us like family … in a good way! Fiona, Cam, and Doug Maki continue to laugh at and endure my antics, and so generously shower us with kindness. Josette and Dr. Geoff Tesson are true grandparents to my kids, and spoil my family and me with their fine cuisine, beverages, and support. They also allow us to tap their maple trees, which is crucial to our annual production of the golden nectar of the woods. Kath Salidas and Norm Dube have become cherished friends and have been extraordinarily big boosters of our kids. Likewise, Ross McKague and Mel Bertrand and their kids, Simon and

Isabel, have become dear friends and help make K-street the best hood in which to live. Dr. Robert "Bob" Hall has been remarkably supportive of our family, practically from the moment we first met, and provides an astute editorial eye when he proofreads my work: thanks Bob. I must also express my gratitude to Jeff Baron (for the walks, talks, runs, and adventures), Sebastian Nault (for the bush expeditions), Bruce Hennessy (for welcoming us to Sudbury many years ago and being a great neighbour ever since), and Patricia Hennessy (for the food, fun, and coaching my swimming and canoeing techniques; may she be comforted by knowing that my record-slow breaststroke across Ramsey Lake in the Beaton was not a reflection on the quality of her mentoring). More recently, Robert "Bob" Narozanski has become a great friend of our family and is a master craftsman when it comes to home renos; he is also extremely patient in mentoring his most challenging apprentice. Finally, Jim Farrell did a superb job of reviewing the original manuscript and providing a wide range of feedback. His keen insights into all things forestry are always appreciated, as is his razor sharp wit. Dr. Sandy Smith, forest entomologist extraordinaire at the University of Toronto, has been a great friend over the years and provided ideas about how to improve the manuscript.

My gang at Laurentian's gym – when it was open, back in the good ole pre-COVID days – gifts me practically every morning with a nonpareil network of true buddies; they comprise my "community" at work. They are genuinely interested in my triumphs and losses, come out to watch my kids' sports, and share my insatiable appetite for appreciating how wonderful life is. The gang's ring leaders include Dr. Stan Holloway, Adel Zaher, Mike Grace, Dr. Sandy Knox, Al Gancher, Dr. Craig Hamilton, Roman Wereszczynsky (aka "Big Guy"), Dr. Ed Najgebauer, Dave Tomini, and Wayne Tonelli.

Then there is my immediate family. My mother chose to live south of Vancouver after she retired, which made it much more difficult to visit her. Nevertheless, her new location afforded my family and me a wonderful place to visit, and it served as a superb base for researching the story of aerial dusting in British Columbia. Many times she took care of my young kids as I took the pre-dawn bus to the archives in Vancouver or researched in Victoria, and in the process she spoiled them with an infinite supply of Fudgee-Os and love. My mother died as I was putting the finishing touches on this book, and although it was the hardest loss I have ever suffered, she was ready to go; it was her time. Nevertheless, I miss her every day and am saddened to know that she will no longer share life's joys with us. However, I see her essence in the beauty of the trees, birds, and animals I encounter as I walk or run in the bush with Lumi, our wonder dog. I am also comforted by the knowledge that she is at peace knowing that her little boy ended up doing okay. I want to thank my wife, Cindy, for continuing to offer me boundless love and support, and many good laughs to boot. How she does so even though we have been together for over a quarter century remains one of life's greatest mysteries. I will be forever amazed how she can still present a look of genuine interest as I babble on about my latest research discovery or project. Nolan and Carling, our kids, grew from infants into amazing young adults as I was preparing this book. I am grateful to them for enriching my life in myriad ways (most of them positive), and look forward to watching them on their future journeys. It is to them that I dedicate this book. To avoid any hint of favouritism, may they know that the order in which their names appear in the dedication does not reflect their standing in my eyes.

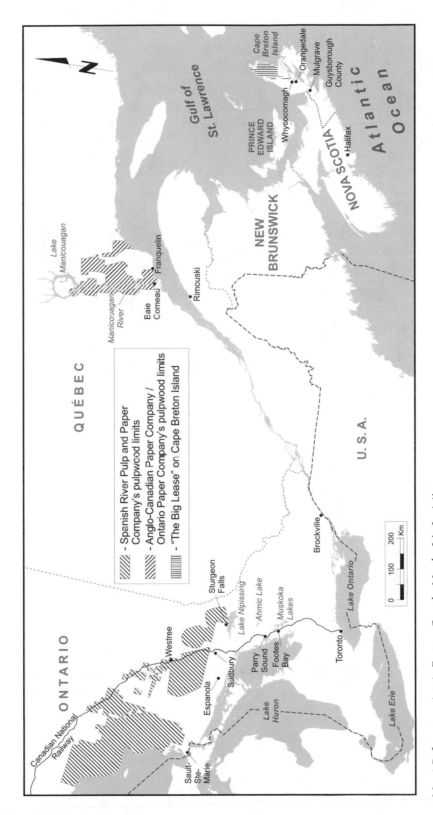

Map 1 Reference points in Eastern Canada. Map by Léo Larivière.

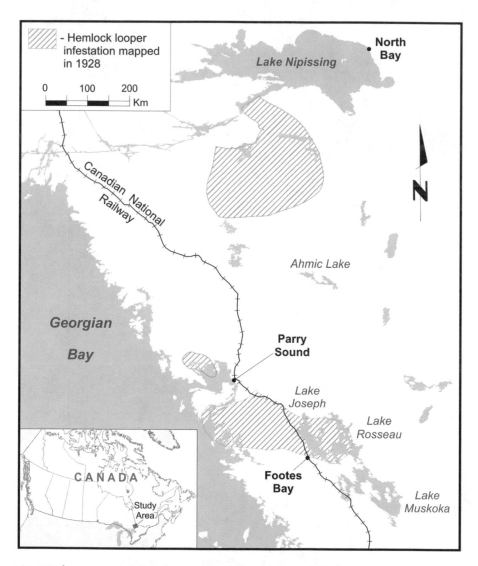

Map 2 Reference points in Muskoka and Parry Sound Districts. Map by Léo Larivière. Source: Information about the looper infestations transcribed from AO, RG1–256, Box 1, Hemlock Looper (2), ca. 1928, loose map in file.

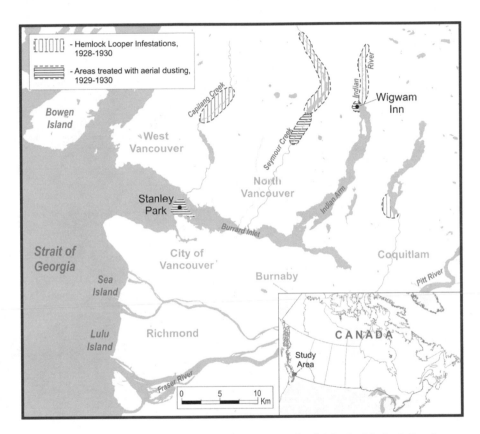

Map 3 Reference points in British Columbia's Lower Mainland. Map by Léo Larivière. Source: Information regarding looper infestations from G.R. Hopping, "The Western Hemlock Looper" (MA thesis, Iowa State College, 1931), 72. Courtesy of University Library, Iowa State University of Science and Technology.

Killing Bugs for Business and Beauty

Canada's Aerial War against Forest Pests, 1913–1930

"The natural question is what can be done to destroy them?"

Something was amiss in Muskoka, Ontario, one of North America's most cherished and exclusive recreational paradises, in the summer of 1927, and authorities were at a loss as to how to deal with it. The area's beloved hemlock trees, which were an integral part of the region's natural allure, were turning brown and dropping their needles at an alarming rate, and the local cottage and lodge owners and summer vacationers were growing more anxious by the minute as their previously picturesque landscape was being ruined. Referring to the bugs eating his trees, one distraught vacationer summed up the general feeling of despair with this twist of irony: "The natural question is what can be done to destroy them?"[1] Provincial forestry officials were summoned to investigate and identified a tiny insect, the hemlock looper, as the culprit. The problem, they openly admitted, was that they had no means of tackling the menace. It was "not physically possible," they declared in throwing up their hands, to kill bugs in forests.[2] Ground sprayers, which were used in orchards to douse infested fruit trees with insecticide, were useless when it came to dealing with harmful insects in tall, difficult to access woodlands.

Within a few short years, however, a handful of government and industry officials in Canada had effected a veritable revolution in the fight against forest insects. They realized this goal after conducting a series of trials in which specially adapted aircraft bombed infested woodlands with chemical dusts (chemical sprays would be introduced much later) that were toxic to the pests. Although the new aerial dusting technology was far from perfect (and those involved in this endeavour realized its limitations), upon completing these efforts, Canada was recognized internationally as being in the vanguard of employing this technique as an effective tool in managing the insects that were destroying valuable parts of the nation's forests. In fact, in the early 1930s the Canadian government published circulars that boasted that it had practically perfected this control method in combatting outbreaks of a few of the country's most common forest pests.[3]

Curiously, the aerial dusting projects occurred to assist two forest users whose interests are most often seen as being diametric. Predictably, they were conducted to help Canada's forest industry, specifically its pulp and paper makers. They conceived of trees as harvestable commodities that they sought to protect until such time as they could cut them down and process them into various commercial products. But the aerial dusting efforts were also undertaken at the behest of those who viewed the trees collectively as an aesthetic commodity that only retained its value if it remained standing and in a healthy state. These people came from all walks of life, although admittedly those whose voices echo through the pages of this book occupied society's upper and middle classes. They were cottage and resort owners, municipal officials and city boosters, members of the media, scientists, and urbanites of all sorts. To a person, they shared a common view of the woods: the trees were beautiful sights to behold and worth preserving. For simplicity, these persons will be referred to as environmental preservationists and recreationists. Although the tree cutters and the tree lovers are often seen as anathema to each other, during the period from 1927 to 1930, both groups looked to aerial dusting as a potential solution to their forest insect problems, and they did so literally *a mari usque ad mare*. Projects were carried out for the sake of the forest industry in eastern Nova Scotia (1927), north of Sudbury, Ontario (1928–9), and in the Baie Comeau region of southcentral Quebec (1929) (see Map 1). They were also conducted at the behest of environmental preservationists and recreationists in Muskoka, Ontario (1928–9) (see Map 2), and in North Vancouver (1929), Stanley Park (1930), and Seymour Canyon (1930) in British Columbia (see Map 3).

In retelling the various parts of this story, numerous themes wind their way through them and tie them together. Predictably, "nature," or the non-human world, plays a central role in any story about bugs and the forests in which they live, and in this tale nature was both an actor and the stage. As an actor, nature demonstrated its agency in several ways. For example, entomologists began realizing very early in their studies the implications of human disturbance in the woods. These scientists recognized that the manner in which industry traditionally cut its timber was creating conditions that were ripe for the populations of destructive forest pests to explode. This included leaving the ground littered with slash and debris in which certain insects thrive. It also entailed harvesting in a manner that promoted the reproduction of tree species that were particularly vulnerable to insect attacks.

In addition, the bugs' different life cycles and habits created unique opportunities and challenges for forest entomologists. These divergent characteristics made it easier to kill some pests and much more difficult to kill others.

Nature also provided the stage for this history of aerial dusting in Canada because so much of this story concerns human conceptions and ideas about wilderness; the projects involved Canadians and Americans alike, and thus demonstrate the prevalence of these attitudes. Those who drove these pioneering efforts obviously sought as their main goal to protect their trees from infestations of forest pests. In doing so, they frequently resorted to the rhetoric of war in an effort to portray their endeavours in terms that resonated with a population that had all too recently emerged from the world's most gruesome conflict. Furthermore, these same individuals often exploited to their advantage widely accepted opinions about these insects and notions of what forests and trees represented, principally by fomenting a crisis that demanded immediate and radical action. While this dynamic was sometimes visible in the editorial cartoons of the day, it always pervaded the discourse the recreationists employed to foster support from government officials to treat their infested trees.

Far more importantly, the evidence that follows demonstrates that these same environmental preservationists – and indeed nearly all citizens – viewed nature through a unique prism that allowed for a most curious form of "selective seeing." When they stared out at the forest, their brains processed what their eyes saw in a way that perceived of the landscape as being pristine. They thus gushingly worshipped the forest for its wilderness qualities, seemingly unable to discern the many signs of human impact on these environments. In Muskoka, these human intrusions ranged from grandiose cottages to immaculately polished wooden boats, whereas in British Columbia they spanned the gamut from a luxurious lodge to watersheds whose lakes and rivers had been completely reconfigured by hydrological experts. Given this distorted perception, it is unsurprising that practically all those involved in these dusting projects, but particularly the environmental preservationists, were blind to the collateral damage these efforts could cause. They could not appreciate that there might be something fundamentally dangerous about applying a deadly chemical to kill bugs in the same forest that provided them with the songbirds and animals they so loved, or the clean drinking water they so dearly cherished.

Another major strand that permeates this story is the success the Canadian scientists and dusting pilots enjoyed and the conditions that allowed them to do so. As the pages that follow lay plain, developing the means by which to drop chemical insecticides over a forest insect infestation in an efficient and precise way was a Herculean task in the context of the 1920s. Nevertheless, Canadian entomologists made remarkable progress in developing this particular branch of science and technology. This was due in large part to their ingenuity and their ability to innovate to solve problems, particularly the practical ones, and in this regard they were true pioneers. They were repeatedly forced to improvise to overcome the seemingly never-ending string of hurdles that came their way. To be sure, they seemed to delight in doing so, developing novel means of stickhandling around the intellectual and practical obstacles that stood between them and success. They were trailblazers who were imbued with a sense of adventure and confidence that they could raise their game to new heights – literally. Equally important was the remarkable degree of freedom they enjoyed from political interference to address the challenges they faced in carrying out their work. So, too, was the courage and skill that a handful of pilots demonstrated as they conducted these novel bombing runs against pests in the woodlands. Despite their lack of training and the profound dangers they faced in flying these sorties, those who undertook them did so with such prowess that observers who witnessed their talents and fortitude in action consistently described their work in heroic terms.

Politics – in terms of both elected officials and government employees – is also a central part of this story in many different ways. At its highest level, Canada's aerial war against forest pests between 1927 and 1930 involved international relations between Canada and the United States. It both underscores the influence Americans wielded over natural resource management policy in Canada during this period, and showcases Canadians' strong inclination to please their southern neighbours. This tale of aerial dusting also sheds light on intergovernmental and intragovernmental relations, and relations between government and the private sector. In doing so, it uncovers a period of truly remarkable cooperation. This was a time during which bureaucratic officials within different branches of government at the municipal, provincial, and dominion levels were able to work hand in hand with each other, and with officials in industry, to accomplish a great deal with precious few resources. Moreover, they were able to do so, and to resolve amicably any conflicts that arose along the way, despite – and perhaps because of – the absence of official contracts that laid out their respective obligations and duties.

Furthermore, the aerial dusting projects of the interwar years reveal a great deal about the nexus of power that shapes environmental policy in Canada, particularly the central role played by pressure politics and lobbying. Undeniably, the scientists who were involved in these pioneering efforts to kill forest insects enjoyed extraordinary latitude in terms of how they conducted their campaigns and the specific insects that the projects targeted. Nevertheless, it was the politicians – and those closely connected to them – who largely dictated for whom and where the missions were carried out. In this regard, access to the corridors of power in Canada deter-mined the location of nearly all of these particular killing fields. At a time when there was essentially no group that opposed the use of chemicals to combat pests, this "inside" lobbying, as it has been termed, was practically all that was needed to win the day.[4] As a result, this story serves as a powerful reminder of the degree to which elite interests unduly influence the government's administration of the pub-lic's natural resources.

Finally, there is much to be learned from reviewing the attitudes and mindsets of the entomologists and foresters who were front and centre in these groundbreak-ing chemical bombing trials against forest insects. Very quickly after they began studying the bugs that were attacking our most valued trees, they demonstrated a remarkable grasp of the ecology of the woods. They understood that eradicating the pests was a chimerical notion, and that these critters would always be present. They also came to understand that human activity was often creating conditions that were exacerbating the problems the insects were creating, and that altering our behaviour was the only long-term solution. Never did these scientists and foresters seize upon aerial dusting, as exciting and glamorous as it was, as the panacea for combatting forest insects.

Ultimately, this book's argument can be distilled into a few essential observa-tions and one overarching message. It reveals the remarkably similar demands that both industrialists and environmental preservationists/recreationists make on the environment. Both strive to manipulate nature out of self-interest, and poli-tics often plays a central role in realizing their ends. Yet, this story highlights that there is one major difference between the two groups. The cottagers and nature lov-ers often endeavour to enshroud their ambitions in terms that profess a profound affection for the non-human environment in an attempt to strengthen their case by depicting it as a moral issue instead of a financial one. Occasionally, overly anxious property owners may stray from this script by inadvertently expressing their true motivations: namely, that protecting nature was all about dollars and cents. These slip-ups matter not, however. The evidence presented here illustrates how broadly society believes that the mission to "save" nature was and is unquestionably a noble one. Furthermore, this moral appeal transcended class differences and justified using public funds to foot the bill for aerial dusting, even though often only a scant few at the top of the socio-economic ladder would benefit from the expenditure. The result is that environmental preservationists typically have an extraordinary advantage in influencing environmental policy because what they seek to achieve – preserving nature's aesthetics – is deemed to be a goal the value of which is inesti-mable and inherently good. In contrast, the tree harvesters have no such argument available to them. Canada's pioneering aerial war against forest pests thus demonstrates how shared impulses often drive both the harvesters and preservers of the trees, and highlights the acute dangers inherent in allowing emotional appeals, rather than logic, to shape environmental policy.[5]

This book is hardly the first to explore humankind's complex conceptions of and relationships with the "natural" world, and as a result it borrows from and builds upon the existing literature. In terms of North America, there exists a long list of authors who have explored one of the book's central subjects: namely, how the continent's citizens have appreciated and defined the aesthetic appeal of the untouched woods, lakes, and mountains around them, and how this concept has been an ever-evolving social construct. A frequently cited passage from Ralph Andrews succinctly captures this phenomenon. Referring over a half-century ago to the mammoth trees that once dominated the Pacific Coast, he writes that "there was beauty, yes … but who would know until men found the forest and judged it so."[6] Likewise, Aldo Leopold, "the father of American wildlife ecology," noted our inclination to be drawn to and value most deeply those aspects of nature that we find pleasing to the eye. As he neatly summarizes it, "our ability to perceive the quality in nature begins, as in art, with the pretty."[7] These "untouched" landscapes collectively came to be referred to as "wilderness," and William Cronon adeptly analyses the term's remarkable history and dramatic evolution over the centuries. He points out how, by the mid- to late 1800s, "wilderness" had come to refer to those sublime, uncultivated places whose physical appearance left their beholders feeling as though they were in the presence of a higher power.[8] Not surprisingly, many environmental historians have traced the fight to protect these sacred sites from being sullied by development.[9]

Also at the core of this book about killing bugs is the story of the various conflicts that complicate how we have worshiped wilderness, and a substantial body of largely American literature explores how this has not been a simple love affair. Leopold was arguably one of the first to identify what he noted as a central paradox that marks our attitude towards wilderness. Although we profess a profound and unbridled adoration for the landscapes that we consider to personify natural beauty because we conceive of them as being untouched by human hands, we have shown an irresistible proclivity to obtrude into these areas in an effort to improve them, with the result that we leave them indelibly altered. As Leopold explains, albeit hyperbolically, "all conservation of wildness is self-defeating, for to cherish we must see and fondle, and when enough have seen and fondled, there is no wilderness left to cherish." In this regard, he levels particularly harsh criticism at those who choose to spend their leisure time camping and canoeing through the wilds. "Barring love and war," Leopold asserts, "few enterprises are undertaken with such abandon, or by such diverse individuals, or with so paradoxical a mixture of appetite and altruism, as that group of avocations known as outdoor recreation."[10]

More recently, a number of authors have focused on how our blindness to the many ways in which we reconfigure nature has inclined us to continue doing so without ever recognizing the significance of our actions; this subject is a leitmotif in the aerial dusting projects described in this book. Daniel Schneider and Richard White are but two of the many American academics who examine this concept. In the process, they identify the resulting "organic machines" and "hybrid nature," particularly in industrial settings, and the degree to which these landscapes are pervaded by contradictions.[11] Similarly, "natural constructivism" is the term that has been coined for the human penchant to worship nature for being untouched and our concomitant irresistible inclination to improve upon it in ways that disguise our influence.[12] Finally, Cronon adds another dimension to this discourse when he describes, to cite the title of his landmark article, "The Trouble With Wilderness." Our tendency to feel as though we are addressing our environmental challenges by preserving sanctuaries of nature that are separate from the quotidian environments in which we live and work is profoundly problematic, he argues, because wilderness thus "represents the false hope of an escape from reality." If these wild sites are the only ones in which we believe nature is safe, and humans must be absent from them in order for the flora and fauna to be healthy, then this places humans outside of nature and bodes very poorly for our hope to resolve the environmental issues that confront us today.[13]

Although Canadian authors have been late to the environmental history party, they are now making up for lost time, particularly in terms of delving into how nature has been conceptualized in this country; in doing so, their views have resonated loudly with similar studies that have been conducted in the United States.[14] Patricia Jasen, for example, surveys the cultural foundations of tourism in Ontario, particularly the fascination with wilderness experiences in places such as Muskoka, and Alan MacEachern investigates how national parks in Atlantic Canada were selected in the twentieth century based upon cultural constructions of what constituted natural beauty. Similarly, Clare Campbell describes the ways in which people's ideas about nature shaped their reactions to Georgian Bay, a renowned vacation spot in central Ontario that abuts Muskoka. She also analyses how "the Bay" region conversely influenced people's conceptions of it over the course of several centuries. Tina Loo's work on the regulation of Canada's wildlife in the twentieth century echoes this theme, particularly the often contradictory attitude taken towards managing nature whereby people destroyed what they ostensibly loved. Her work also underscores how the elite in Canada fought for the state to embrace and police its conception of nature. Likewise, Sean Kheraj touches upon many of these same ideas in his work on Stanley Park, a locale that plays a significant role in the story that this book tells. Finally, Joy Parr provides highly original insight into how we experience nature through our five senses.[15]

There is also an extensive literature that deals with another major topic that this book addresses; namely, the history of science in North America in general, and entomology in particular, the insects that have threatened the continent's food and forest crops, and the measures that have been taken to combat them. Among others, Suzanne Zeller paints the broader context for the story by describing how leading officials in Canada during and after the Victorian period demonstrated a deep and abiding faith in science because they saw it as representing the magical key that would unlock the industrial potential of the country's enormous natural wealth. Michel Girard builds on this subject by tracing the history of Canada's Commission of Conservation (1909–21), and James Whorton highlights this fervent confidence in science and technology to explain the paucity of concern that was expressed by the general public about the use of chemical insecticides in the United States in the years leading up to the Second World War. Rachel Carson's epic *Silent Spring* traces the story over the next few decades, and in the process she points out the irony in our inclination to poison our food in order to save it. Thomas Dunlap surveys the rise of DDT (dichloro-diphenyl-trichloroethane) as the miracle insecticide in the mid-1940s and the subsequent battle among scientists, citizens, and the public policymakers that led to it ultimately being banned. Lastly, Edmund Russell is the best known among several authors to have chronicled how the strategies, technologies, and metaphors of war shaped the approaches to and understandings of insect control in the United States during the twentieth century. Again, all these themes are woven into the story that is told here.[16]

Furthermore, there is a relatively large volume of literature specifically about the history of Canadian entomology and the country's war on bugs. Stephane Castonguay has made a number of crucial contributions to it. He chronicles the birth and growth of economic entomology – the science of controlling harmful insects in commercial agriculture and forestry – in the country from the late 1800s until 1959. During these years, he argues, the ranks of the federal entomology service grew by leaps and bounds, and its primary role changed from disseminating knowledge to conducting scientific experiments into both the life cycles of the pests that were wreaking havoc on Canada's food and tree crops and the best measures for combatting them.[17] George M. Cook traces the introduction of chemical insecticides suitable for use in fruit-growing and farming operations in Canada in the late nineteenth and early twentieth centuries, and he argues that the diffusion of this technology during these years, and the granting of legal authority for such activities, "provided the technical and legal basis for rapid expansion of the institutions of applied entomology in the years after 1910."[18] Clinton L. Evans and Paul W. Riegert examine the use of herbicides and insecticides against agricultural pests in Western Canada from the late 1800s until roughly the end of the Second World War, and Richard Rajala focuses on federal forest entomology in British Columbia in general and the Vernon laboratory in particular during the 1900s. A few senior officials from both the provincial and federal entomological services in Canada have also authored histories of their jurisdictions' attempts to investigate and control forest insects, and the official biographer of the Ontario Provincial Air Service notes its involvement in aerial dusting during the late 1920s.[19]

Finally, *Killing Bugs* is also positioned at the intersection of business and environmental history, where it engages the large body of literature that analyses lobbying, pressure politics, and public relations in North America. Long before Canadian officials conducted their first aerial dusting project in the 1920s, lobbying was a well-established business practice. Canadian authors such as Donald Creighton and Michael Bliss, for instance, trace how myriad groups presented their cases to colonial officials in an effort to secure the grant, monopoly, or concession they desired in New France and British North America.[20] Similarly, historians such as Ben Forster, Tom Traves, Ken Cruikshank, and Arthur Lower assess how, just before and after Confederation, varied interests brought pressure to bear on political officials in an effort to win advantages such as favourable tariffs and freight rates and access to valuable resources.[21] A number of studies have also investigated how competing interests battle it out to influence environmental policy, in particular during the modern era. Douglas Macdonald explores the subject across many industries in the national context, while Mark McLaughlin, L. Anders Sandberg, and Peter Clancy examine how this dynamic played out with regard to aerial spraying of forests pests in the Maritimes.[22]

Despite the impressive nature of the foregoing articles, monographs, and reports, these studies leave several gaps that this book endeavours to fill. For example, a number of previous works on the history of Canada's struggle to combat forest insects mention the projects that occurred in the period under examination here, but do so in only a cursory manner.[23] Most importantly, the in-house government histories provide what are largely narratives without any significant analysis of the events. As a result, none explains the dynamics that led to the realization of the aerial dusting projects between 1927 and 1930. In particular, they ignore the major role that nature lovers played in coating some of their most treasured wooded retreats with toxins in an effort to save them.

The format for this book is entirely geographic and largely chronological. Chapter 1 traces the origins of entomology in Canada, its rise as a branch of the dominion government's bureaucracy, and how its growth was fueled by the search for means to combat pests that were menacing the country's food and forest crops. The military technology developed during the First World War raised the possibility that aircraft could help realize this end, but the means were fraught with challenges. The first chapter takes the story up to the mid-1920s and in the process highlights how, from the very beginning, the effort to kill forest pests aimed to realize aesthetic goals as much as industrial ones. Chapter 2 tells the astounding tale of the inaugural dusting experiment, which occurred in Nova Scotia in 1927. Indeed, the very fact that it took place at all, no less in a Maritime province better known for its fish than its forests, is rather startling. It only happened because of a remarkable alignment of goals among a number of disparate groups and individuals. These included the dominion entomologists, Nova Scotia's newly crowned Chief Forester, well-connected American business interests, and a native Bluenoser politician who wielded considerable power at the national level. Chapter 3 brings the story to central Canada, specifically the aerial dusting projects that took place in Ontario and Quebec over the course of 1928–9 to assist some of the leading forest companies there. These projects underscore the impressive level of cooperation that marked all the aerial dusting projects during this period, as well as the industry's faith in the power of science to resolve many of the challenges that it faced. In addition, nature continued to flex its muscles and influence events on several levels in this campaign against forest pests, and those involved in the effort continually did the same through their ingenuity and resourcefulness.

The second half of this book examines the projects that were carried out to protect the aesthetic, rather than industrial, value of Canada's forests. In doing so, its main feature is the paradox that defined the environmental preservationists' drive to protect the parts of nature that they loved by killing those elements of it that they despised. This theme is central to chapter 4, as are several of those outlined in the preceding paragraphs. It focuses on the aerial dusting work that was done in Muskoka in 1928–9 to satisfy the local recreationists, whose prized hemlock trees were being attacked by a native pest. Although they stressed that this effort was essentially a public service because the stunning local environment was invaluable, their motivation was undeniably driven by self-interest. Moreover, this chapter underscores the singular power of a handful of upper class urbanites to influence the government's management of its woodlands, and the potentially catastrophic

perils – such as poisoning the local water supply – that can result from this process. These themes are also prevalent in chapters 5 and 6, which focus on British
Columbia. Chapter 5 traces the story of Canada's war on forest bugs back to the
early 1910s when entomologists first attempted to kill forest pests with chemicals
shot from ground sprayers in Vancouver's Stanley Park. The next operation was
carried out in BC a decade and a half later, this time using aircraft and targeting a
stand of infested trees near a plush lodge that was a short ferry ride from downtown
Vancouver. BC's remoteness from the Canadian entomologists working in Ottawa
meant that the handful of officials in BC who conducted the aerial dusting projects
in that province did so virtually on their own. All these threads also weave their
way through chapter 6, which investigates the operations that occurred in BC in
1930, namely in Seymour Canyon and Stanley Park. Once again, the story was all
about killing bugs that lurked in these forests in order to preserve their aesthetic
qualities.

The conclusion does more than merely review the case that the preceding chapters present. It reinforces the fact that Canada enjoyed significant success in conducting these aerial dusting experiments in the years before the Great Depression
and was thus recognized as a world leader in developing this new technology. At
the same time, it highlights how the country's entomologists still faced problems
that they needed to iron out. Perhaps more importantly, it dives even deeper into
the many contradictory ways in which Canadians and Americans conceived of the
woods for which they espoused such a profound love. This includes attaching different values to trees depending upon their location, the use to which humans put
them, whether they were sick and dying trees, and the circumstances surrounding
their demise. In addition, this story highlights how those who were involved in
these projects were so focused on the potential benefits of destroying the target
insects that they were blind to the collateral damage the chemicals caused in the
woods. Unfortunately, this myopia was a chronic condition and had major implications for future efforts of this type. Finally, the conclusion highlights the preponderant influence urban elites wield over our country's environmental policy and the
many pitfalls of allowing them to do so.

CHAPTER ONE

"Airplane dusting offers the only present hope"

PREPARING TO TAKE CANADA'S CAMPAIGN AGAINST
FOREST INSECTS TO THE SKY, 1884–1926

For as long as humans have inhabited North America, they have been sharing the continent with insects. Some of these creatures have drawn rapt attention because they were perceived as being lovely marvels of nature. In this regard, the monarch butterfly, with its graceful, seemingly effortless flight pattern and regal tawny and black colouring, and the dragonfly, with its iridescent body and filigree wing structure, spring instantly to mind. At the other end of the spectrum are the "pests," whose ranks are populated by a seemingly endless stream of bugs that suck our blood, bite off chunks of our flesh, and devour the food and forest crops that we value so highly. While Canada's Indigenous peoples undoubtedly admired the lovely six-legged creatures, they took measures – such as using fire and smoke – to defend against the pests that tormented them and attacked their resources. The waves of European settlers arriving in Canada likewise encountered these same pests and reacted similarly.[1]

The situation in Canada began changing during the mid-nineteenth century when the formal study of insects began. A number of factors contributed to the establishment and early development of entomology as a discrete field in Canada, not least of which was a profound faith in the power of science to solve the problems that the country was facing. In the process of investigating the pests that were attacking Canada's forests, entomologists came to realize that the way in which humans were harvesting trees was ironically creating conditions that were ideal for the propagation of those very same bugs. Consequently, they concluded that the only long-term solution to this problem was to alter how loggers were cutting the woodlands. In the short term, however, the scientists sought a means by which they could effect "direct control" of insect infestations.[2]

One solution presented itself immediately after the First World War, but it was hardly a quick fix. Recently developed technologies presented the possibility of using aircraft to bomb the forest pests with deadly chemicals. The campaign to realize this goal received an added boost from the use of war rhetoric, a tactic that helped the idea gain traction among the general public. Nevertheless, significant hurdles stood in the way of this campaign achieving success. In particular, all the principals involved recognized very early on that conducting this type of operation was both perilous to the pilot who flew it and inordinately expensive for its sponsors.

All the while, there was a very curious twist to this effort to tackle the bugs that were feasting on Canada's trees. As much as this agenda was driven by the desire to assist the country's timber industry, it was seen just as much as a means of protecting the aesthetic qualities of Canada's forests. Humans conceived of the insects' rapacious ways as destroying the charm of the woods. As a result, using deadly toxins to eradicate the pests lurking amidst the trees was viewed as a way of "saving" the natural beauty of the woodlands. Remarkably, barely a word was uttered about the hypocrisy inherent in expressing a deep appreciation and love for nature on the one hand, and using those same words to justify killing it on the other.

Entomology, or the study of insects, was one of the first sciences developed by humans, and with good reason. Biting bugs collectively represent at best an annoyance and at worst a major health hazard because of their ability to transmit disease. Moreover, early civilizations largely depended upon the cultivation of crops for survival, and insects loomed as potentially menacing threats to the production of foodstuffs. Farmers grappled with these quandaries in ancient times, and in doing so helped spawn the field of entomology. The challenges faced in this realm increased exponentially as agriculture modernized and intensified during the eighteenth and especially into the nineteenth centuries. The clearing of huge swaths of forest cover to create farmers' fields often exacerbated the problem by destroying the habitat within which predators of the "pests" lived. Raising monoculture crops on ever greater tracts of land, which was, for example, the story of prairie settlement in Canada during the late 1800s and early 1900s, created conditions ideal to fostering epidemic outbreaks of insects that feasted on these plants. Developments in the field of transportation – both in terms of infrastructure (i.e., canals and railways) and better and faster modes of moving goods – worsened the situation by introducing foreign insect invaders to areas in which they could destroy plants without fear of being subject to their natural enemies.[3]

These conditions in Canada gave rise to the field of agricultural entomology, albeit later than much of the Western world. Initially, amateurs studied British North America's farm insects. Lacking specialized, formal training in a classroom setting, these self-taught naturalists became experts in their field through years of first-hand observation. In an effort to increase the profile of their work and foster support for it, three dozen such amateurs joined forces to form the Entomological Society of Canada in 1862, which has been described as "Canada's oldest continuing 'national' specialized scientific society."[4] It took a few decades, however, for the dominion government to begin establishing a bureaucracy dedicated to carrying out entomological work; one of the first major steps it took in this direction was bringing James Fletcher into the fold. He was a former banker and then employee of the Library of Parliament who had dedicated his leisure time both to reading as much as he could about insects and examining them in the field. In turn, he had also freely offered advice about pests to elected officials, whose respect for his knowledge was reflected in their decision to appoint him the country's honorary entomologist in 1884. Within a few years, the dominion government copied the template the Americans had laid out by establishing a network of experimental farms across the country through which it created a Division of Entomology and Botany.[5] Fletcher was officially appointed Dominion Entomologist (and Botanist) in 1885. Over the next few decades he tirelessly devoted himself to his work, an effort that included establishing an "insect intelligence service" across the country. It involved amateur entomologists – from backyard gardeners to commercial fruit growers – forwarding their observations about the bugs they encountered to Fletcher. His biographers cite his profound commitment to furthering the entomological cause in Canada as the reason for his premature death.[6]

Beginning in the waning years of the nineteenth century, several forces significantly raised entomology's profile within the ranks of the dominion bureaucracy. In 1899, Fletcher tapped Arthur Gibson of Toronto to be the dominion's Assistant Entomologist. Gibson was only twenty-five years old, but his work as a lepidopterist – a specialist in the study of butterflies and moths – had attracted Fletcher's attention and earned Gibson the job opportunity in Ottawa.[7] Together, the two men nudged their work into the modern era. Initially, they focused largely on creating an inventory of the noxious insects that were deleteriously affecting Canada's most important food crops, specifically wheat grown in the Prairies and the soft fruits raised in orchards in places such as Nova Scotia's Annapolis Valley and the Niagara region of southern Ontario. Under Fletcher and Gibson, the Division of Entomology focused increasingly on studying the life cycles of problematic bugs in an effort to understand how to better control them, and devoting more resources to publicizing its findings.[8] This evolution was expedited by the discovery of a foreign invader, the brown tail moth, on nursery stock shipments imported to Nova Scotia in the early 1900s. The ease and speed with which a tiny insect could wreak havoc on one of the country's most profitable agricultural industries spurred the dominion government to dedicate far more resources to its entomological work. In the wake of the brown tail moth's arrival, Ottawa created a discrete Division of Entomology in the Department of Agriculture to address it and other pest problems that were afflicting the country. The government appointed Dr. Charles Gordon Hewitt to head up the new program as the Dominion Entomologist upon Fletcher's death in 1908.[9]

Although Hewitt served only a relatively short term in this post – he died of pneumonia in 1920 – his tenure was formative for Canadian entomology. He identified how two factors – the behaviour of the pests he and his colleagues were studying and the country's seminal political document, the *British North America Act* – could be conceived of in a way that created a logical division of labour for Canada's entomologists. From an insect's perspective, there were no provincial boundaries. Bugs could flit and crawl from Alberta into Saskatchewan without even noticing the time change. Many provinces were thus afflicted with infestations of the same insect, and Hewitt recognized that it would be best to prevent each province from duplicating research into these shared problems. Moreover, Canada's original constitution allocated control over natural resources to the provinces. As a result, Hewitt recognized that it made sense to have a national agency oversee and coordinate entomological research and have the provinces be responsible for applying control measures in the field.[10] In fact, this already seemed to be the de facto arrangement when he had arrived on the scene, but his presence confirmed that it would be the framework through which future work would be carried out.[11]

Hewitt's arrival was important for another reason as well. The Canadian Division of Entomology had headhunted him from a university in Manchester, England, where he had been working as a lecturer in economic zoology after earning three degrees – including a doctorate – in that field from the same institution; Castonguay describes Hewitt as having been "a British entomologist on the rise."[12] He thus represented the first professionally trained scientist within the Division of Entomology in Canada, and his hiring reflected the degree to which university-educated specialists would thereafter dominate this field. Hewitt ensured this would be the case by aggressively engineering a dramatic expansion in his organization's scope of activities and personnel ranks. This included shifting his division's focus from public relations and outreach work to scientific research, whereby insect infestations were accepted as permanent features of the environment requiring constant vigilance and study to address. Hewitt gained greater authority and independence for his activities in 1914 when officials in Ottawa agreed to carve out his branch from that which dealt with the experimental farms. He was also highly successful in terms of luring to Canada several of North America's foremost entomologists, some of whom will be mentioned in the pages that follow. He did so in an effort to staff the twelve new laboratories he established across the country and the four new departmental branches he created under his authority. The latter dealt with naming and identifying insects (systemic entomology), economic entomology and control (i.e., field-crop and garden insects), plant quarantine, and forest insects.[13]

Work in the last field – forest entomology – had actually been going on for some time both in Canada and around the world. In Germany, for instance, where forestry was a well-developed science by the early 1800s, insects frequently posed a serious and pressing threat to the trees that were critical to the forest industry; learning about these pests thus became a subject of significant importance to woodland managers. Such concerns were still far off for those engaged in harvesting timber in Canada during the early to mid-nineteenth century, however. At this time, the industry was concentrated in eastern Canada, and it relied upon cutting white – and smaller volumes of red – pine, species that were as of yet unaffected by either disease or pestilence.[14] But when the pulp and paper industry arrived in Canada in the late 1800s, suddenly insects became a major concern to those whose business relied upon harvesting timber. The new mills that began springing up – chiefly in Quebec and Ontario – were drawn by several forces, one of the most important of which was a seemingly infinite supply of spruce and balsam fir trees. These species were certainly found in abundance in eastern Canada, but they were also vulnerable to a few natural enemies.[15]

The most important of these was the misnamed spruce budworm (*Choristoneura fumiferana* Clemens). It is a native pest that, in eastern Canada, feasts primarily on balsam fir, and to a less extent on white, red, and black spruce. The adult insect is a small moth whose wings are mottled in brownish hues, and who, during July and August, lays its apple-green eggs in clusters of from 10 to 150 on fir and spruce needles. The caterpillars hatch from the eggs in about a week and immediately begin spinning their cocoons on the underside of the branches; they spend their winters in these silken cases. The following spring, as the buds of their host trees open, the caterpillars emerge from their shelters and feed upon the new, succulent foliage for roughly one month. The caterpillars then spin new cocoons and metamorphose into pupae. Roughly ten days later the moths appear and begin the process anew. The injury the host trees suffer is cumulative, and each successive attack reduces growth and new foliage. The trees can survive if the infestation lapses after a few years, but most will die if it lasts much longer.[16]

Back in the early 1900s, Hewitt was most concerned about this novel bug. He discovered a budworm infestation in the Gatineau River valley about 100 miles north of Ottawa in 1909. When the infestation expanded dramatically the following season, he sounded the alarm bells about the toll it was taking on the area's balsam and spruce trees.[17] With the Canadian pulp and paper industry growing exponentially at this time, it is no surprise that Hewitt's news brought unprecedented attention to the field of forest entomology. No sooner had he identified this menace than he and Arthur Gibson, the dominion's Assistant Entomologist, had begun investigating the budworm's habits and life cycle. In 1911, Hewitt dramatically raised the profile of the study of insects that were damaging Canada's woodlands by hiring James Malcolm Swaine as the Officer-in-Charge of the recently created Forest Insect Section of the Division of Entomology. Swaine arrived as a highly qualified and well-respected scientist and would go on to enjoy a lengthy career in the division in particular, and the dominion civil service in general. He was born in 1878 in Barrington, Nova Scotia, and he began working as a schoolteacher and then attended the Nova Scotia Agricultural College. He went on to earn both his BSc in Agriculture and MSc from Cornell University in 1905 and 1906 respectively. Thereafter, he taught Entomology and Zoology at Cornell and then at McGill's Macdonald College in Montreal before being hired by the dominion government. His speciality was bark beetles, and he prepared a major report on them for the Division of Entomology in 1917, the research for which he also used towards earning his PhD from Cornell two years later.[18]

Under Swaine's guidance and with Canada's forest industries generally prospering, forest entomology grew slowly but surely. Swaine's expertise in bark beetles drew him westward where these bugs were a major concern to the lumber producers in British Columbia, whose business enjoyed a major boost with the opening of the Panama Canal in 1914. To help address their concerns about the threat that beetles posed to their timber stocks, Swaine oversaw the establishment of a laboratory in the province and the hiring of another university-trained entomologist, Reginald C. Treherne, on the eve of the First World War. With Hewitt and Gibson at this same time conducting studies of the spruce budworm back east, forest entomology was laying a firm foundation for itself in Canada.[19]

But Swaine and his colleagues still faced an uphill battle in terms of establishing their field of forest entomology in Canada. Professional forestry had only arrived in the country around the turn of the twentieth century, and Canada's first forestry school – based at the University of Toronto – had been created in 1907. At this time, the nation could count the number of practising foresters on both hands and still have fingers left over. Although their focus was fixed firmly on managing woodlands to maximize the efficiency with which timber could be harvested over the long term, in the public mind the practice of forestry became practically synonymous with tackling what was then seen as the greatest threat to Canada's woodlands: fire. The damage it caused to the landscape and the toll it took in human lives were staggering. Stories about the havoc it wrought in hinterland communities – authors are still writing about the Miramichi River conflagration of 1825[20] – made it public enemy number one as far as forests were concerned. Canada's pioneer foresters thus devoted much of their time to improving the extent to which they were able to prevent fires in the regions under their administration. Within this context, the damage insects were causing in the woods seemed inconsequential. No human lives had ever been lost or settlements razed because of tiny insects defoliating trees or boring through their bark. As a result, among the relatively small group of Canadians who were worried about acting as responsible stewards of the country's forests in the years before the First World War, forest entomologists were the black sheep.[21]

Acutely aware of their "second-class standing," forest entomologists took advantage of every opportunity to upgrade their status. They waged a tireless public relations campaign to convince Canadians of the bona fide peril that insects posed to the health of the country's trees and the industries dependent upon them. In fact, bug scientists repeatedly pointed out that the evidence they were gathering indicated that forest pests were destroying, albeit in an insidious fashion, far larger swaths of Canada's woodlands than fires. Typical of their messages was the one that Hewitt delivered in a bulletin Swaine authored about bark beetles for the Entomological Branch in 1918. In his foreword to the document, Hewitt emphasized the enormous threat that beetles posed to the country's coniferous forests. "Forest fires are spectacular," Hewitt stressed, "and the results are immediately and strikingly noticeable, but competent authorities are of the opinion that the annual loss caused by the depredations of these and other forest insects which are widely distributed throughout the country is greater in the aggregate than the loss due to forest fires."[22]

Demonstrating the utility of entomological work was a much more forceful way to raise the field's profile, and so the ultimate goal for both forest and agricultural entomologists became developing methods for controlling the harmful pests that they were identifying. By the time war had broken out in Europe in 1914, Canada's agricultural entomologists had developed a range of strategies for dealing with problematic insects. On the most basic level, they employed manual methods. To prevent the importation of exotic insects, for example, the government created a system for funnelling all plants and nursery stock entering the country through one of a series of inspection stations that were set up on or near Canada's borders. To deal with bugs – both foreign and domestic – once they had already established a beachhead in the country, agricultural entomologists hired teams to inspect each potentially infested plant or orchard tree and destroy any signs of harmful insect activity. This approach could certainly prove effective, but it was both labour intensive and immensely time-consuming. It also became less practical as Canadian farming increasingly came to resemble modern industrial agriculture. These factors help explain why scientists across much of the Western world had begun experimenting during the late 1800s and early 1900s with applying toxic chemicals – with unassuming names such as "London purple" and "Paris green" – to field crops and fruit trees that were suffering from harmful infestations. Canada's agricultural entomologists followed suit. No sooner had Fletcher been appointed Dominion Entomologist in 1884 than he begun peppering his annual reports with recommendations of which first generation insecticides to apply to kill various garden and orchard pests.[23]

But the introduction of these toxins to the war on those bugs that were eating food crops was hardly a panacea to the problem. It certainly represented a major improvement over the existing methods of insect control. Dousing infested plants and orchard trees with a chemical solution was definitely easier and more efficient than removing the pests by hand, and the effectiveness of delivering the toxic sprays was greatly improved when more powerful hand- and gas-powered pumps, often mounted on wagons, were introduced. But this method of dealing with the insects' ravages presented its own set of practical challenges. Most pests were only vulnerable to poisons for very short windows during their life cycles, specifically when they were gorging themselves on their host plants, and they were most prone to the toxins when they were relatively small. Because spraying was still a leaden and tedious process, however, the poison could only be applied to a very limited number of affected plants while the target insects were most susceptible to it. Moreover, the treatment could only be applied to plants that grew either close to the ground or were very easy to access, and the chemicals were expensive.

These factors meant that it was highly inefficient and uneconomic to treat vast swaths of food crops, such as wheat, whose value was relatively low compared to the area of land upon which they grew. In contrast, the high value of the produce grown in orchards, and the relative ease with which the trees could be treated, made them the primary targets of the inaugural spraying campaigns in Canada at the turn of the twentieth century. Even in these instances, however, administering these treatments was very expensive, and better application techniques were still eagerly sought.[24]

Forest entomologists faced even greater practical challenges in their quest to control harmful insects. Whereas farmers' fields and orchards were often bordered by roads and could be reached easily by foot or horse, the opposite was true of Canada's commercial woodlands. They exemplified the remote nature of much of the country's treed landscape. Moreover, this was an era during which practically the only means of moving logs from the bush to the mills was by driving them down streams and rivers and across lakes; there were almost no forest access roads at this time. In addition, insect infestations in the woods could cover tens or hundreds of square miles. And even if workers could have been brought into these remote areas to treat an insect epidemic manually, trees grow exponentially higher than most food crops. It was thus immensely difficult and costly to climb up a forest tree to clip the infested branches so that they could be destroyed, and it was virtually impossible to reach them with ground sprayers.[25]

But then the First World War, specifically all the technological innovations it spurred, opened up enormous possibilities for both farm and forest entomologists to overcome many of the challenges they had heretofore faced. Many fields of endeavour benefited from the advances made during the conflict in Europe, and aeronautics received one of the greatest boosts. Aircraft – both fixed-wing and dirigibles – repeatedly proved their value as instruments of war. While the mud and trenches made movement on the ground remarkably dangerous and exhausting, planes and blimps were able to fly high above the engulfing quagmire and hail of bullets to provide a bird's eye perspective from which to monitor troop activities on the front. In addition, aircraft could inflict direct harm on the enemy by dropping rudimentary bombs and strafing its soldiers. The introduction of poison gas to kill troops en masse also created another potential tool for entomologists. They saw in this horrific development a way for deadly chemicals to be unleashed in vast quantities to fight problematic insects.[26]

After the war, it quickly became apparent that aircraft could be immensely helpful in a wide range of peace-time pursuits, and this was especially true for those who worked in Canada's forests. Prior to this time, the only way to survey a tract of trees was to paddle along its waterways and then traipse on foot through an obstacle course of fallen trees and broken branches, impassable swamps, and rocky crevices. Aircraft would revolutionize the process by which foresters investigated the state of the woods they managed by allowing them to create timber inventories and maps in a fraction of the time such work had previously taken.[27] They could also use planes to patrol for forest fires, and to transport personnel and equipment to fight the blazes. Furthermore, the development of planes that could take off and land on the water rendered them even more useful for this work in the Canadian hinterland, where open stretches of flat land are scarce but waterways abound. No sooner had the war ended than contractors began springing up to provide these flying services to interested clients.

Circumstances dictated that the dominion government would also play a prominent role in performing these duties for the forest industry and other sectors of the economy. During the war, over 20,000 Canadian military personnel had served in the Royal Flying Corps (the precursor to the Royal Air Force) and the small, England-based Canadian Air Force that had been created in September 1918. Soon after hostilities ended, the British government gifted the Canadian Department of Militia and Defence (DMD) with 100 planes in recognition of its contribution to the British war effort. The DMD thus had hundreds of trained pilots and a fleet of aircraft at its disposal, and its senior officials realized that the very existence of their fledgling air force – the Canadian Air Force would be officially formed in 1920 and the Royal Canadian Air Force (RCAF) four years later – rested upon their ability to find civilian applications for its services. This was truly a unique situation, wherein this particular branch of the military set out to commercialize its operations in the fullest sense of the word. To realize this end, senior air force officials waged an unabashed public relations campaign to advertise their talents throughout the 1920s. The greater the civilian demand for their planes, they reasoned, the more their budgets and flying fleet would grow. Moreover, they were eager to adapt aircraft for civilian use because they believed doing so would assist them in developing new planes for wartime applications and help foster a domestic aircraft designing and manufacturing industry. The Air Board, with a dominion minister at its head, was created in 1919 to oversee these activities, and it existed until the RCAF was formed to take over its duties five years later. In 1927, the RCAF was relieved of direct responsibility for control over civil aviation and the new Directorate of Civil Government Air Operations (DCGAO) was given authority over this activity (during the 1920s the DCGAO was often referred to as the Air Board, and for simplicity the latter term will be used herein).[28]

The growth of the RCAF during the first decade of the interwar period was a symbiotic process, but military officials certainly led the charge. Their campaign was overseen by Sir Willoughby Gwatkin, the chief of the general staff, but it was energized by two individuals from the Air Board. John A. Wilson served as its secretary during the early 1920s, and in 1927 he was appointed Controller of Civil Aviation when the RCAF separated its civil from its military functions. Similarly, Lieutenant Colonel Ernest W. Stedman was the director of the Air Board's technical branch and then became one of the RCAF's senior engineers. Both exploited every possible opportunity to publicize the tremendous potential of aircraft in all sorts

of work. Fortunately for them, there were also forces outside the military establishment that wished to realize some of the same ends. Senior officials within other
branches of the bureaucracy in Ottawa were keen to employ the dominion government's fleet of aircraft to serve their purposes as well. The thick historical record that
chronicles the air force's civilian activities during the 1920s attests to the degree to
which those who were responsible for administering and developing Canada's vast
natural resources, specifically its bounty of farmland, mineral deposits, and tracts
of timber, eagerly embraced the Air Board's offer to assist them in their work.[29]

In fact, hostilities in Europe had barely ended before the Air Board had begun
dispensing its services back home. It cooperated with the largest timber licensee
in Quebec – the St. Maurice Forest Protective Association – to conduct a series
of groundbreaking projects. They included the first Canadian aerial forest surveys and fire patrols and the transport of officials to remote locations in the bush.
Observers raved about these efforts, and one of Canada's most progressive foresters,
Ellwood Wilson, immediately became an enthusiastic advocate for using planes in
his line of work. He published an article in a leading international aviation journal
that extolled the virtues of employing aircraft for civilian purposes in general and
forestry in particular. To reinforce his case, he described how a forester could get a
much clearer idea of a 50 square mile tract of trees during a two-hour flight than he
could during a two-week trek on the ground.[30]

Forest entomologists were anxious to share in the benefits that this squadron of
government planes could deliver, and they succeeded in arranging for the Air Board
to carry out a project for them at this time. Already in 1919, the forest entomologists had been able to procure the Air Board's services for one of their men, Eric
Hearle. He was conducting a survey of sites in British Columbia's lower mainland
that were contributing to a major mosquito problem. Hearle's aerial photographs of
the insect's breeding areas proved so useful that he was able to persuade his superiors to support his request for greater use of aircraft in entomological work.[31] A
cooperative effort was also undertaken over the course of 1920–1 by the soon-to-be
disbanded Canadian Commission of Conservation, the Air Board, and the dominion's Division of Forest Insects. It focused on mapping a spruce budworm outbreak
that had been festering in Quebec and the Maritimes and was moving west, having already crossed Ontario's eastern inter-provincial border. The aerial reconnaissance operation was based in the Haileybury region of northern Ontario, and the
mapping highlighted how the infestation had extended its reach south along Lake
Temiskaming and northward across the height of land into the Temagami Forest
Reserve (Image 1.1). Arthur Gibson, who had recently been promoted to Dominion
Entomologist after Charles Gordon Hewitt's death in 1920, was astounded by the

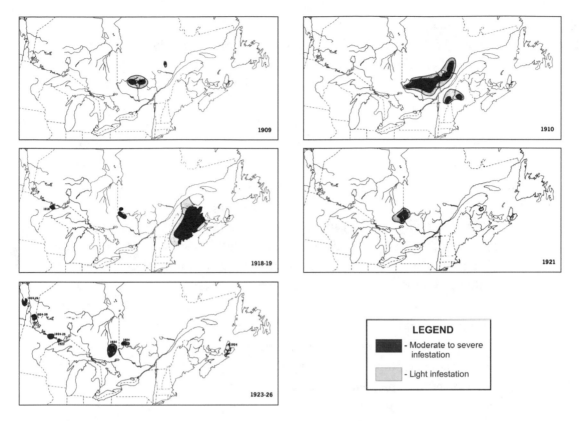

1.1 Evolution of spruce budworm infestations in Eastern Canada, 1909–1926. Source: Adapted from C.E. Brown, *A Cartographic Representation of Spruce Budworm* Choristoneura Fumiferana *(Clem.) Infestation in Eastern Canada, 1909–1966* (Canadian Forest Service Publication 1263, Department of Fisheries and Forestry, 1970). Reproduced with the permission of the Department of Natural Resources Canada, 2022.

efficacy with which the aircraft could be used in this type of work. "The information obtained from a few days' flying," he gleefully exclaimed in his annual report, "would have taken two men more than six months to acquire by ground surveys."[32] Canada's entomologists had truly broken new ground in successfully conducting this latest undertaking. Even the Americans would for years thereafter recognize it as having represented the first major effort to survey and map an insect infestation from the air.[33]

Several factors ensured that Canada's cadre of forest entomologists and officials from the Air Board would have plenty to discuss during the early 1920s. The former were inspired to employ more planes in their work by the extensive use their colleagues in the Division of Botany had been making of the Air Board's services in studying the problem of white pine blister rust in British Columbia. The botanists openly admitted that they never would have known about the extent of the disease without the Air Board's help. Moreover, the forest entomologists were dealing with a series of ongoing insect infestations. The aforementioned spruce budworm outbreak in eastern Canada remained active. Although it died back in New Brunswick over the early 1920s, by the middle of the decade new infestations were being tracked in northeastern and northwestern Ontario and northwestern Quebec's Rouyn district, and on Nova Scotia's Cape Breton Island (Image 1.1). Furthermore, the Canadian forest entomologists were monitoring infestations in both the east and west of the tent caterpillar (*Malacosoma disstria* Hbn.). It was devouring sugar maple in Quebec and aspen in Saskatchewan and Manitoba.[34]

The tent caterpillar also became a marquee attraction in the campaign the dominion government's entomologists launched at this time to elevate significantly the profile of their activities and improve the effectiveness of their efforts, and Swaine was in the vanguard of this movement. In the spring of 1923, the Entomological Branch began publishing *The Canadian Insect Pest Review* to promote its activities and highlight the largely ignored danger that insects posed to the nation's farms and forests. In addition, the journal sought to help coordinate the work Canadian entomologists were doing during this period and served as an immensely helpful supplement to the literature that the provincial entomological groups published during these years. Notably, during the *Review*'s first few years of existence, it focused on the destruction that the tent caterpillar was causing in various areas across the nation and stressed the urgent need to combat it.[35] It was thus hardly surprising that this insect served as the first potential target when Swaine recommended experimenting with chemicals dropped from the air to combat forest insects in Canada. Although Arthur Gibson, the Dominion Entomologist, had as early as December 1922 suggested that this approach be taken by distributing, as he put it, "poisoned dust from [a] hydro or aeroplane,"[36] Gibson was simply conveying Swaine's wishes to the Department of National Defence (DND). A few months later, Swaine discussed the idea with officials from the Air Board and requested that they arrange for at least one of their planes to "sprinkle poison dust" on the tent caterpillar outbreaks in Quebec and Saskatchewan.[37] Although Swaine continued to insist that the best way to deal with forest insects over the long term entailed changing the industry's approach to harvesting, he nonetheless remained, for at least the next decade, the undisputed champion of the Canadian campaign to use aircraft to drop chemical toxins to kill these pests.[38]

While it may have been predictable that the tent caterpillar would be chosen as the target insect for the inaugural aerial dusting trial, the pre-eminent reason for selecting the site to be treated was a bit surprising. By early 1923, Swaine had been providing updates for a few seasons on the tent caterpillar's depredations across Canada. In the east the outbreak had now waned, but the infestation in Saskatchewan was both ramping up and predicted to continue intensifying, and so attention focused on addressing this menace post-haste. It was centred on the Moose Mountain Forest Reserve in the southeastern section of the province. Amidst the virtually treeless prairie plains, the Reserve consisted of a 160-square-mile oasis of poplar, birch, and ash forest that the Dominion Forest Branch had set aside in 1908. It had done so because the timber tract had become the local area's de facto woodlot and was in need of protection and management if it were going to continue playing this crucial role far into the future. In addition, the park's "beautiful scenery" and its most important water body, Kenosee Lake, had begun drawing locals who wished to enjoy their warm weather vacations there. The forest reserve had thus been created as much for its recreational allure as its worth as a source of timber. Significantly, when it came to justifying why the Reserve's tent caterpillar infestation warranted treatment, Swaine emphasized that it was "used as a summer resort for people of southern Saskatchewan and the defoliation by the caterpillars practically destroys its value for the time."[39]

Although a mission of sorts would be conducted to combat this pest, the Air Board was not eager to undertake it. The Forest Insect Division cooperated with the Dominion Forest Branch in the spring of 1923 to conduct a ground-spraying operation using lead arsenate against the tent caterpillar in the Moose Mountain Forest Reserve, but they were hampered by practical limitations. William R. Motherwell, Canada's Minister of Agriculture, explained that they could only douse "trees adjoining roads and open spaces" with the poison, which greatly limited the effort's effectiveness.[40] This gave Swaine even more reason to pressure the Air Board into conducting its inaugural aerial dusting campaign against forest insects in the reserve, one that Saskatchewan's minister of agriculture actively supported. But the Air Board made it clear that it had two major reasons for rejecting these entreaties. First, it lacked a "hopper," the receptacle that stored the chemical dust on the aircraft and from which the dust was released as the plane flew over the infested forest. Second, and much more importantly, the Air Board felt that the work was far too dangerous for it to take on at this time. It simply lacked a plane that was capable of skimming just above the forest canopy and maneuvering safely around any protruding tree crowns. In fact, this serious hazard was one that Swaine and his fellow forest entomologists also recognized at this time.[41] As a result, even though Swaine was already working with the Air Board on developing a Canadian hopper, at this point the notion of killing forest insects by bombing them with toxins was an uninviting one to those officials who had authority over the country's air force.[42]

The RCAF's top brass would soon be facing an even stronger lobby from Swaine after developments south of the border significantly strengthened his hand. The United States had been a world leader in adapting aircraft for civilian uses. In the wake of the First World War, several American scientists had begun experimenting with using former military airplanes and blimps to apply chemical poisons to trees and cultivated crops in an effort to eradicate both destructive weeds and insects. In August 1921, for instance, the American government's Aviation Experimental Station at McCook Field in Dayton, Ohio, was the base for a novel trial. It saw Lieutenant John A. Macready fly a former warplane, a Curtis JN-6, which "dusted" a nearby and very small (6-acre) grove of catalpa trees with lead arsenate in an effort to kill the larvae of the catalpa sphinx (*Ceratomia catalpa* Bvd.). In all, the effort dumped roughly 175 pounds of toxic dust on the trees, and it proved to be a smashing success. The scientists who carried it out published an article in *National Geographic Magazine* that raved about their results. "The outstanding feature of the application," they proudly declared, "was the remarkable precision with which the poison could be placed at the point intended, thus dispelling the idea expressed by many before the test was made that the poison dust would be tossed willy-nilly by the air currents – wholly beyond control." Moreover, the toxin had had a lethal effect on the caterpillars. Walking under the trees a few days after they had been treated, the authors of the article described how "not a step could be taken without crushing numbers of [the insects], some of which already had begun to putrefy. ... The effect on the insects had far exceeded our fondest expectations. ... A careful investigation revealed the astonishing fact that not over 1 per cent of the caterpillars remained alive on the trees, and the minute observations and notes by the experts who witnessed the test preclude the idea that the destruction of the insects could be attributed to any other agency than the poison."[43]

News of this success spread like wildfire through the aviation and mainstream media, and it unleashed calls for this technology to be broadly applied in the battle against bugs. One potential application involved tackling a highly destructive pest in the Deep South. The United States Bureau of Entomology had established the "Delta laboratory" in the early 1900s in Tallulah, Louisiana, to conduct research into insects that were menacing southern field crops. Foremost among them was the boll weevil, which was ravaging Dixie's most important agricultural product: cotton.[44] After first studying the weevil's life cycle and behaviour, personnel at the Delta laboratory in Tallulah began conducting experiments during the First World War that used chemicals to kill the pest. Trials demonstrated that calcium arsenate, a dry powder, was remarkably effective in this regard, but applying the toxic dust proved problematic. The entomologists in Tallulah first used hand sprayers and then mules to broadcast the chemicals. These beasts of burden initially carried perforated bags of poison dust that were spread among the plants as they lumbered through the fields, and later they pulled wheeled dusters through the crops while a machine operator dispersed the dust. All these methods proved immensely inefficient for treating large areas, however, and were labour intensive and involved cumbersome equipment. They were also practically impossible to use after it rained.[45]

In the wake of the successful aerial dusting project in Ohio in 1921, Dr. Bert R. Coad, the American entomologist who was leading the effort to eradicate the boll weevil in the Cotton Belt, was inspired to take his campaign of "insect warfare," as one article termed it, to a higher level.[46] Initially, he employed army-loaned aircraft and pilots to develop the means for delivering the poison from the planes to the fields, and the enormous advantages of applying the dust in this manner were immediately apparent. The aircraft could treat relatively vast fields of crops with tons of the calcium arsenate dust in a fraction of the time that it would have taken ground crews to perform the same task. Coad's laboratory also carried out an experimental campaign against mosquitoes by treating flooded forest areas, and its results were equally impressive. Only a few seasons of testing convinced Coad of the efficacy of this delivery system and the tremendous promise it held for all sorts of agricultural and forestry applications. To realize this potential, he cooperated with a plane manufacturer, the Huff-Daland Company of Ogdensburg, New York, to design and begin producing a specialized aircraft to perform this unique type of work. By 1925, the Huff-Daland "Puffer" was dropping tons of calcium arsenate over cotton fields in Dixie, and Coad and others were authoring reports that beamed with the details of the revolutionary breakthrough they had made in the war against destructive insects. The firm soon became known as the Keystone Aircraft Corporation, and so its dusting aircraft was commonly called the Keystone Puffer.[47]

In fact, Coad and his colleagues made such significant progress that they received a resounding endorsement of their work, which generated even more publicity for it. The Board of Governors of the National Aeronautic Association of the United States passed a resolution at its meeting in July 1925 stating that "the use of airplanes for the extermination of pests and the protection of agriculture has proved successful."[48] No sooner had it done so than the Canadian media began broadcasting news of the Association's declaration. Editorials began calling on officials in the dominion to undertake dusting trials against a host of insects that were attacking Canada's food and forest crops.

Swaine seized upon the progress the Americans were making as fodder in his campaign to help Canada's fight against forest insects take flight. In fact, his

American allies had already done yeoman service north of the border in terms of propagandizing their efforts. In late 1921, for instance, one of the entomologists who had overseen the experimental dusting of the catalpa grove in Ohio earlier that year delivered a paper in Toronto at the annual gathering of the Association of Economic Entomologists. It described in glowing terms the project's success and the potential to utilize this approach in solving a series of insect problems. Thereafter, interest in aerial dusting grew in Canada, and Swaine got wind of Coad's trials in the southern United States. He promptly arranged to see them for himself, trekking down to the Delta laboratory in Louisiana in November 1924. After witnessing the projects that Coad was supervising there, Swaine returned to Canada more determined than ever to introduce aircraft to his work. He thus immediately ramped up his lobby for his superiors in the Department of Agriculture and the DND to support this initiative. To Joseph H. Grisdale, the Deputy Minister of Agriculture, Swaine stressed that recent experiments in the United States had proven the remarkable effectiveness of aerial dusting, and that Swaine had already cooperated with the Canadian Air Board in developing a hopper for conducting this type of work in the dominion. At the same time, Swaine admitted that there were still several unknowns in terms of using planes to dust forests, most importantly the danger and cost. As far as the former was concerned, however, Swaine pointed out that great improvements had already been made: the latest planes had much greater reserve engine power, and padding and support in the cockpit. These advances led him to believe that the newer models of aircraft could, if needed, crash land in the forest "with little danger to the pilot and not very much to the machine." He accepted that the expense was still the limiting factor, but that the upcoming summer would be an ideal time to investigate the cost of carrying out a dusting campaign in a Canadian woodland.[49]

Swaine leaned particularly heavily on the DND, and did so in a most shrewd manner; he cunningly planted grandiose notions of the coup it would be for the fledgling RCAF to become a world leader in the aerial war against pests. Writing to John Wilson, Secretary of the Canadian Air Board, in December 1924, Swaine explained that there were a number of tree stands ready for dusting across Canada, including the forest tent caterpillar infestation at Moose Mountain in Saskatchewan and the spruce budworm outbreak in northeastern Ontario. Swaine stressed that he favoured treating the latter because "airplane dusting offers the only present hope for direct and rapid control of outbreaks such as that of the Spruce Budworm. The great gain that would result from success warrants making a considerable effort in experiment." He added that any new aircraft that the RCAF would have to buy for bombing the pests could be adapted to many types of aerial work, such that it could divide the total cost of the plane among several users. Swaine then under-scored the nub of the matter. He recounted the enormous progress the Americans had made in this field but that "no actual dusting work on forest land has yet been carried out in a practical way for defoliating insects. ... The possibility of success seems sufficiently hopeful that the immense savings that could be made if success were achieved warrants, at least, a very careful consideration."[50]

For several reasons, officials from the RCAF swallowed the bait that Swaine offered hook, line, and sinker. They were profoundly committed to demonstrating the utility of their work to all Canadians, and here was a golden opportunity for them to do so; the allure of spearheading an innovative facet of aerial technology was also too tempting to resist. John Wilson, the RCAF's secretary, arranged for Swaine to meet with the RCAF's director and one of its senior pilots to discuss the matter and to chart a path forward. In late January 1925, Wilson confirmed to Swaine that the air force was investigating whether it would be necessary to pur-chase a Keystone Puffer for the dusting work or if the RCAF could convert one of the planes it already owned for this purpose. The latter option would be much more cost effective, Wilson explained, because it would allow the air force to standardize its equipment and minimize maintenance costs. "The importance of your propos-als is fully recognized," Wilson assured Swaine, "and if this method of combatting forest insect pests can be developed, it might be of incalculable value to the forest industries of the country. We are therefore anxious to co-operate in so far as lies in our power." He closed by asking Swaine to provide a detailed outline of the aerial dusting work Swaine hoped that the RCAF would perform that season.[51]

Perhaps hopeful that he was about to reel in his trophy fish, Swaine responded to Wilson's request for more information by including a description of aerial dust-ing's potential in Canada's forests that grossly embellished its possibilities. In terms of his wish list for the RCAF's services in the summer of 1925, Swaine described how conditions in the Moose Mountain Forest Reserve made it a relatively easy site to treat. In contrast, he added that his field staff had not yet identified an area in the budworm infestation in northeastern Ontario that was suitable for dust-ing, but that it made more sense to address this problem first because it was "the injury we are most anxious to control by this method." Swaine then sketched out the scale of the project he desired, basing his figures on the work that the Delta lab had performed in the southern US. In his mind, there was little distinction between aerial dusting over cotton fields and aerial dusting over coniferous for-ests. Because the Americans had covered as many as 1,000 acres an hour in their operations, he postulated that it was "safe to estimate a square mile [i.e., 640 acres] an hour of actual flying time for forest dusting. The cost of the dust will be so high that it is unlikely we shall be able to afford to dust more than ten square miles in this experiment."[52] Parenthetically, although Swaine's calculations were no doubt fuelled by his genuine enthusiasm for the novel experiments he was on the cusp of undertaking, unbeknownst to him, the forest dusting projects that he and others would ultimately oversee in Canada over the next half decade would treat a *total* area of under thirteen square miles.

Joseph Grisdale, the Deputy Minister of Agriculture, was convinced that Swaine's faith in the efficacy of aerial dusting Canada's forests was well placed, and so Grisdale buttressed Swaine's case with the RCAF in February 1925. In doing so, Grisdale pressured it to allocate the resources needed to acquire an aircraft to carry out aerial dusting experiments. His message reminded the air force officials of the smashing success the Americans were enjoying with their dusting program, and that the new type of specially adapted aircraft had drastically reduced the risk of accidents. Grisdale also emphasized that dusting by aircraft was "the only conceivable method of offering any hope for the direct control of great outbreaks of Spruce Budworm and insects which work in a similar way." He committed his department to purchasing the calcium arsenate needed for the dusting experiments if the RCAF would provide the plane, and closed by underscoring that that spring was the ideal time to undertake this initiative.[53]

But then Swaine's campaign suffered a major setback. By this time, several agencies had published reports on the dusting work that the Americans were doing at the Delta lab. Although the story they told was one of a practically unqualified triumph, and rightly so, a discerning eye that read between the lines could discover the potentially perilous nature of this work. In November 1924, for example, the Georgia State College of Agriculture published a booklet that broadcast the remarkable progress the cotton dusting trials were making and how the college was now participating in them. The document traced how the Keystone Puffer had been specifically designed to possess the characteristics needed for this type of work. In particular, it had to have excellent maneuverability and control, especially at slow speeds, and it had to be able to climb rapidly. Moreover, "the aircraft must be rugged and simple … [and have] … the ability to stand up under forced landing and protect the pilot in case of serious crash." This last feature was essential, because the plane was frequently forced to make emergency landings in "cotton fields and tall grass or grain."[54] Bert Coad, who was in charge of the Delta lab in Louisiana, had authored a similar report on the dusting work in January 1924 in which he brought this danger element into much clearer focus. In a thorough evaluation of all aspects of the airplane trials, Coad stressed that aerial dusting could only be conducted over flat terrain, and that it would be "too dangerous" to perform this work over most of the Cotton Belt because it was too hilly. "Commercial use of the planes," the document declared, "could never be developed if the flying should prove unduly dangerous to the pilots, but under the conditions prevailing in these districts low flying is not seriously dangerous. Forced landings can be made at any point, and even though the motor should be cut off while flying over the fields, if sufficient speed were maintained to bring the plane to a level, it could be landed in the corn or cotton with very little danger to the pilot, although of course some damage would be done to the plane."[55]

Here was the Achilles heel in Swaine's campaign to bring aerial dusting to Canada's forests. He had repeatedly tried to gloss over the danger inherent in this type of work. His sanguine portrayal of the Puffer's potential for dusting forests had downplayed – cynics would say ignored – the gaping chasm that separated flying the plane over trees in a "natural" forest from piloting it over farm fields in the southern US. True, the Americans had conducted their groundbreaking experiment against a "forest" insect in Ohio in 1921, but the grove of catalpa trees that had been dusted was located in the middle of a wide open swath of flat farmland. The same was essentially true of the cotton fields in the Deep South. Both represented relatively level terrain with practically no obstacles around which the Puffer had to maneuver.[56] The photographs of these sites told the story better than any words in any report ever could (Images 1.2 and 1.3).

As a result, when the RCAF offered its sober second thoughts on the proposed forest dusting project in March 1925, the safety of its pilots was one of its two major concerns. Basil Hobbs, an RCAF Squadron Leader, had actually travelled to Ogdensburg, New York, to witness for himself the Puffer's abilities. Although he was impressed by the plane's capabilities, he was adamant that it not be flown over treed areas. Soon thereafter, the RCAF's senior officials informed Swaine and his superiors that, in their collective opinion, "forest dusting against insect pests could best be carried out … by using a 'blimp' or 'lighter than air' ship. While pilots often had to take the risk of flying low over forest for a limited time, [w]e did not consider it wise to take on any work which entailed continuous flying in heavier than air craft immediately over tree tops. Sooner or later a serious accident would occur. The 'blimp' was therefore the solution, though the capital and maintenance would be higher than for an aeroplane or flying boat."[57] The problem then boiled down to one of dollars. George J. Desbarats, the DND's Deputy Minister, explained to his counterpart in the Department of Agriculture that "[t]he requirements of the work to which we are already committed will make it impossible for the Air Force to purchase a special aeroplane for this work during 1925, as the funds allocated to civil operations are already fully compromised."[58]

1.2 Aerial dusting of catalpa trees in Ohio, 1921. A Curtis JN-6 aircraft is releasing the poison dust beside the grove of catalpa trees; the photograph was taken from an accompanying plane. The arrow is pointing in the direction the wind was blowing, which wafted the dust onto the trees. Source: *National Geographic Magazine*, March 1922, 338 (Image 605151).

In the meantime, the potential of chemicals to combat destructive insects in both farmers' fields and foresters' woodlands was gaining increasing attention, particularly among the politicians and senior bureaucrats in Ottawa who were inclined to share their views with the RCAF. Over the spring of 1925, the latter received a clipping from a leading Toronto daily that described in glowing terms the remarkable work that was being done in dusting peach orchards in Georgia. Around the same time, the DND also found amidst its incoming mail numerous inquiries from American aerial dusting companies that were eagerly offering to conduct the experimental projects that might be planned for north of the border.[59]

William Motherwell, the Minister of Agriculture, was particularly affected by the publicity campaign being waged by entomologists in general and aerial dusting advocates in particular. He had been deeply moved by the paper that Dr. Clell L. Metcalf, an American entomology professor, had delivered in early 1925 at a meeting of entomologists in Guelph, Ontario. Entitled "Warfare Against the

1.3 Aerial dusting of cotton crops in Georgia, 1924. This sequence of photos illustrates the terrain over which the aerial dusting of cotton crops occurred. The plane is a Huff-Daland Model Thirty-One, which was very similar to the one that Canadian entomologists would use in their forest dusting projects. Source: G.B. Post, "Boll Weevil Control by Airplane," *Georgia State College of Agriculture Extension Division Bulletin 301* 13, no. 4 (November 1924), 6.

Insects," Metcalf had underscored the extent of the bug problem facing humanity. He highlighted how insects were often fighting for the same resources as humans, and that the former had the advantage because they had established themselves on earth eons before the latter. He described in gruesome terms how insects attacked humans and animals, and estimated that the total cost of the damage so far was in the range of $1.5 billion, a truly astronomical figure in an era during which "billion" was a term that was still used only sparingly. These factors, Metcalf argued, left entomologists no choice but to develop means by which they could control bugs, or insects would threaten our very existence. Although he called for Canada to fight a multi-pronged war against the pests, he stressed that chemicals would play a central role in this conflict, and that the Americans were rapidly perfecting the means by which airplanes could apply them.[60]

Having just come through the unimaginably gruelling Great War, such calls to arms resonated with Canadians, and Motherwell was ready to lead the charge in this latest conflict.[61] In an impassioned address in the House of Commons in May 1925, he invoked memories of the battles which had just been fought when he predicted that "the next great war" would be conducted using chemicals against insects, and implored his fellow citizens to recognise the dire nature of the situation.[62] A few months later, Canada's national news magazine, *Maclean's*, used similar imagery in publicizing the entomologists' campaign to the entire country in a lengthy article entitled, "The Greatest War of All." The hyperbolic piece stressed that "all the world knows of Canada's bitter sacrifice and splendid accomplishments in the last great war ... yet even Canadians know little of a far greater war which their government is waging today. It is a war against a power more colossal, more implacable than Germany ever dreamed of being; a war that has been raging for years past and will flame for years to come." It depicted this enemy as launching its invasion across the entire country and fighting with a "relentless, appalling ferocity." The pests attacked everywhere: underground, on the earth's surface, and in the air. Fortunately, Canadians had a leader in defending against the insects, namely, Arthur Gibson, Dominion Entomologist. "To say that he and his assistants are engaged in a colossal war is more than a figure of speech – it is grim, undeniable fact," the article declared. Here was a blatant attempt to engrave in the public consciousness the need both to fear insects and employ military technology to combat them.[63] Grateful for the magazine's support, in early September Motherwell wrote J. Vernon MacKenzie, the general editor at *Maclean's*, to thank him. "I have felt for some considerable time," Motherwell explained, "that the Canadian people generally did not realize the magnitude of the work that confronted our Entomological Branch and their war against all destructive insect life. I feel that [in your] write-up you have given prominence that will do much in the way of making it easier for our men to carry their regulations out."[64]

Buttressed by these shows of support, the dominion's forest entomologists regrouped in 1926 to lobby the DND to make one of its aircraft available for dusting work, but again to no avail. They repeated many of their previous arguments for gaining the RCAF's support, including the "inestimable value" that would result from demonstrating that outbreaks of spruce budworm could be controlled by aerial dusting.[65] The forest entomologists now considered this insect to be their number one enemy, especially because the tent caterpillar infestation in Saskatchewan had subsided. They specifically asked for the RCAF to conduct a ten-day project against the budworm outbreak in either northeastern Ontario or on Cape Breton Island. They hoped that the fact that the RCAF had done no work for the forest entomologists in 1925 would increase the likelihood that the former would grant their request. Again, however, the DND refused to conduct the aerial dusting project to combat the budworm, and in this regard, the forest entomologists were in good company. The RCAF had also rejected the Division of Botany's request for it to carry out an aerial dusting experiment to control grain rust in Manitoba.[66]

The dominion's entomologists realized that they required additional – and ideally creative – quills in their quivers, and they did all they could to find the best ones possible; in doing so, they also demonstrated their remarkable grasp of the ecology of the woods in which the forest companies were operating. In September 1926, Swaine authored an article in the *Forestry Chronicle* that emphasized the potentially enormous threat that the budworm represented to eastern Canada's pulp and paper industry while underscoring how fortunate it was that the country's entomologists had been studying the menace for over a decade. Their investigations, Swaine pointed out, had confirmed that the best long-term solution was for those persons who directed the mills' harvesting operations to reorient their approach. The industry officials who supervised their firm's logging operations had traditionally focused on extracting spruce to the virtual exclusion of all other species because it was considered so valuable for making newsprint, but this approach had depleted the woods of spruce both in the present and also in the future by eliminating the potential seed source for the next crop. It had also created openings in the woods that were ideal for balsam fir, the budworm's favoured feast. The solution, Swaine argued, was to harvest much more fir and much less spruce, thereby diminishing the potential budworm problem.[67]

In the meantime, however, Swaine stressed that there was but one way forward. He contended that the industry would be dealing with periodic epidemics of budworm even if it altered its modus operandi in the woods, and that mechanical control of the insect was "practically impossible." As a result, the most promising solution to the present outbreak of this pest lay in "the distribution of poison dusts" by aircraft, particularly in light of the promising experiments with aerial dusting of cotton insects that were being conducted in the southern United States.[68]

Swaine's campaign received two other boosts. Firstly, senior officials from within the dominion bureaucracy, including the deputy minister from the relatively powerful Department of the Interior, renewed their lobby to push the RCAF to enter the aerial dusting game, specifically against the spruce budworm epidemic on Cape Breton Island in 1927. They stressed that conditions for doing so at this time were propitious because Nova Scotia's Provincial Forester and a company with large pulpwood interests on the island were keen to cooperate in the venture.[69]

While this undoubtedly increased the pressure on the airmen to take action, arguably the most important boost to the campaign at this time was the approach Hans T. Gussow, the Dominion Botanist, took to the matter. He was losing patience with the Air Board's refusal to assist him by conducting aerial dusting experiments to combat wheat rust in the Prairies. At the same time, he was acutely aware of the Air Board's urgent need to demonstrate its value to the Canadian public and its burning desire to be on the leading edge of aeronautical developments in North America. He thus devised a coy means for exploiting these dynamics. He explained to his superiors in the Department of Agriculture that it was exigent to carry out the aerial trials to combat wheat rust, and that this work required a plane that could fly at slow speeds and close to the ground over the grain crops. In other words, he needed "a plane similar to the one in use in the United States [i.e., the Keystone Puffer]." Gussow then played his trump card. He threatened to hire Americans to do this work in Canada if the RCAF was not up to the challenge. "I might mention," Gussow slyly declared, "that United States sulphur interests might be induced to extend their work with planes into Canada, but we would prefer to see any credit resulting from such work go to the Canadian Air Board rather than any United States concerns." Gussow's views were conveyed verbatim to the Air Board.[70]

His appeal to the Air Board's sense of patriotism and pride apparently struck a very sensitive cord with the airmen, and although it effected in them a newfound commitment to undertake the aerial dusting work, the news was not all good on this front. Certainly the Air Board's officials were delighted to learn that, over the course of late 1926, the RCAF was busily investigating various options for planes that would be suitable for conducting the dusting experiments. But in evaluating the different aircraft, the RCAF's personnel decided that a Keystone Puffer – the plane that had been specifically developed in the United States for aerial dusting – was simply too expensive; for the price of *one* Puffer the RCAF could purchase *three* other different types of planes. Moreover, the RCAF still entertained deep concerns about the safety of using any flying machine to apply chemicals to forests. J. Stanley Scott, the RCAF's director, explained to one of his senior colleagues that this type of operation would entail "considerable risk to the personnel undertaking" it. As a result, he recommended first testing this type of project "over the Prairies with special equipment, before undertaking forestry work."[71]

Undaunted, the Entomological Branch renewed its campaign to win over the DND. In December 1926, Arthur Gibson informed J.H. Grisdale, the Deputy Minister of Agriculture, that Swaine was determined to carry out a dusting trial against the spruce budworm the following year. Once again, Gibson emphasized that the infestation on Cape Breton Island presented an ideal opportunity for such a project, particularly because various local interests were eager to cooperate in the venture.[72] Grisdale passed this information, along with the entomologists' wish list for flying operations the next summer, to the DND. He added that the Department of the Interior, a powerful player in the national government's bureaucracy, had now grown deeply concerned about the budworm problem, and that the Deputy Minister of the Interior, William W. Cory, was "prepared to make just as strong recommendations that this work be carried out next year as I am, and, in fact, I am writing to him suggesting that he communicate with you in regard to this matter."[73] Cory dutifully followed this directive, and urged the DND to facilitate a trial dusting against the budworm in Nova Scotia in 1927.[74]

Fortuitously for Swaine and those in the Entomological Branch, recent news from Europe provided them with even more ammunition in their campaign to win over Canadian air force officials. In late 1924, French foresters had detected a limited infestation of pine looper caterpillars, or "fidonie" (*fidonia piniaria*), which were attacking pineries in the Haguenau Forest, France's largest unbroken woodland located near the French–German border. The local authorities had immediately applied traditional methods of fighting the infestation, but they proved ineffective. By the spring of 1926, the size of the outbreak had grown from roughly 40 hectares (nearly 100 acres) to over 900 hectares (roughly 2,200 acres), prompting the Inspector of Forests at Haguenau to call for drastic new measures. Germany had been grappling with an equally serious caterpillar problem, and already in 1925 it had experimented with dropping insecticide dust by plane over a portion of the infected forest. While this was a rare occasion on which the French officials had been eager to follow their neighbours' lead, they were unable to adapt a suitable aircraft and produce the requisite toxic dust until the fall of 1926. They carried out their effort over two days in late October. The lateness of the season meant that the caterpillars were especially large and strong, so the French foresters used dusts containing much higher concentrations of insecticide than the Germans (the French dusted using pharmacolite with 40 per cent arsenical acid instead of the 13 per cent concentration used in Germany). The French dusted a total of 53 hectares (about 130 acres), and the report on the operation noted that its cost was high because this was the first time that the country had carried out this type of project on such a scale.[75]

Despite the high cost, the Inspector of French Waters and Forests declared the operation a smashing success. He praised the pilots for having carried out their missions to "perfection," noting that they had flown at between ten and fifteen metres above the tree tops without incident. In looking to the future, he recommended that subsequent projects of this nature be carried out on a large scale (to minimize their cost) over flat land (to facilitate the dusting) at the most favourable time of the year, and that action should be taken immediately upon detecting future infestations. As the Inspector of Forests at Haguenau concluded, "[t]he flights for dusting powder, executed in my presence absolutely convince me that this operation is quite susceptible to generalization, on account of its rapidity and great efficacy, if it is used at the right time; I consider, therefore, that this particular use of aviation presents an important interest and a real future."[76]

Finally, Swaine and senior officials at the RCAF were driven by another animus to get into the aerial dusting business immediately. Over the spring of 1926, they had corresponded about the best design for a hopper, the receptacle which stored the dust on the aircraft until it was dumped on the insects. Their exchange revealed that Swaine and a few of his colleagues in the Department of Agriculture had actually drawn up blueprints for one a short while earlier, and for more than merely practical reasons. "These were in connection with our efforts to prevent a German firm obtaining a patent in this country," Swaine confided to John Wilson, the RCAF's secretary, "covering the general principle of distributing dust from planes."[77] Swaine was thus keen to conduct a dusting trial as soon as possible, and in the process filed all the necessary paperwork so that he could avoid being forced to buy foreign technology – at a premium – when Canada took its war against pests to the air.[78]

Nevertheless, the DND was still disinclined to commit to carrying out such a project in Canada, and so its senior officials temporized as best they could. They assured Swaine and his colleagues that they were investigating the available options in terms of the type and cost of aircraft, and in fact they were true to their word. They asked their American colleagues – through their British representative at the Air Attaché office in Washington, DC – for as much information as possible about aerial dusting, and their southern neighbours graciously obliged.[79] But still, the guiding lights at the RCAF remained non-committal about conducting an aerial dusting experiment over a tract of Canadian forest.[80]

As a result, Swaine was unable to get his project off the ground as 1926 came to a close, but circumstances would change dramatically early in the new year. To be sure, the Air Board had demonstrated a bona fide interest in carrying out such an endeavour. It was, however, deeply worried – and rightly so – about the safety and cost of conducting this type of operation; the latter issue had been grounds for the Board refusing to purchase the Keystone Puffer. What was lacking in this instance were a few critical factors that gave the Air Board no choice but to enter the business of aerial dusting over woodlands. These forces would coalesce very quickly in 1927 as the year began to unfold. At that time, a few individuals exercised crucial political influence at several levels to launch the dusting projects, specifically in the windswept and often fog-enshrouded forests of Cape Breton Island.

CHAPTER TWO

"One of the first aerial applications of an insecticide in forestry"

THE POLITICS OF BATTLING THE SPRUCE BUDWORM IN NOVA SCOTIA, 1925–1927

Nova Scotia is hardly the first province that springs to mind when one thinks of Canada's forests and its forest industry, but it was the site of the world's first aerial dusting operation against insects in a natural forest.[1] In fact, it hosted not one but two separate aerial dusting projects in mid-1927. To be sure, the pump had been primed. For the better part of a half decade, dominion entomologists had waged a sustained campaign to push Canada's Air Board, and the Department of National Defence (DND) under whose auspices it operated, to experiment with using aircraft to dump poisons on forest and farm pests. For its part, the Air Board was keen to demonstrate its usefulness to all Canadians and to the world moreover, and had endeavoured to cooperate. It had a problem with the means to the end, however, and it repeatedly argued that the risks and costs of this type of work were simply too high.

Just as it seemed the logjam would never be broken, a range of forces at work in Nova Scotia in the mid-1920s changed everything. There, a spruce budworm outbreak centred on Cape Breton Island was discovered within the context of a most bizarre sequence of events occurring in the province's woodlands and corridors of power. Otto Schierbeck, a Danish immigrant and forester in Nova Scotia, was waging an aggressive and, as it will become clear, at times outlandish campaign to demonstrate that province's pressing need to hire him as its first Chief Forester. A central element in this ploy was his portrayal of the budworm outbreak in Nova Scotia's eastern reaches as a calamity of epic proportions, and one for which he alone had devised an ingenious solution. In this effort, Schierbeck was championed by Frank J.D. Barnjum, an outspoken forest industrialist who sought to exploit his political connections to realize his aim of building a large pulp and paper mill in Nova Scotia. Barnjum saw foisting Schierbeck into the position of Chief Forester as a means to achieve his own ends.

These forces alone would not have brought the aerial dusting program to Nova Scotia, for this drama required a few more central characters to realize that aim. Several of them came from the highest ranks of the American government and business community, and one of them was a leading Canadian Member of Parliament. James L. Ralston was a native of Nova Scotia who, as luck would have it, was wielding considerable power in Ottawa by the mid-1920s as Minister of National Defence, the same department that was responsible for Canada's Air Board. In his legal practice, Ralston represented the major American timberland holders in Nova Scotia whose tracts were threatened by the local spruce budworm outbreak. The upshot of all these dynamics interacting was quite remarkable. In the spring of 1927, Nova Scotia played host to the world's inaugural attempt to use poisons dumped from airplanes to control a pest in a natural forest.[2]

This chapter thus highlights several of this book's main threads. The most important one is the cardinal role that politics played in determining how and where the dusting projects occurred in Canada during this period. Actors on the international and national stages certainly played leading roles in shaping events, but those in the provincial arena were also important. In an indirect way, political considerations also ensured that the dusting work that was done in Nova Scotia in 1927 was much more dangerous than it ought to have been. Nature was also active in increasing the hazard involved in this work. The budworm's behaviour and life cycle shaped how the scientists attempted to combat it, and also demonstrated that it would be a very difficult creature to kill using this latest technology. So, too, did the make-up of the forest in which it lived, and the budworm's natural enemies also skewed the results that the scientists derived from their dusting trials in Nova Scotia. Furthermore, the theme of cooperation is also prominent in this chapter, specifically among the wide range of government agencies and industrial players involved, as is the degree to which the Canadians were true pioneers in this field. They continually had no option but to rely upon their own ingenuity to move the dusting projects forward, and their efforts ranged from gauging whether the poison dusts were effective to figuring out how to drop the toxins in a controlled and measurable way.

The history of Nova Scotia's woodlands is unlike any other forested area of Canada. Prior to Euro-Canadian development, the province was reasonably well endowed with a wide variety of trees. It is located within the Acadian Forest Region, which was dominated by several coniferous (e.g., balsam fir, red spruce, and white pine) and deciduous (e.g., sugar maple and yellow birch) species. For thousands of years, the local Indigenous peoples had utilized the trees they needed for purposes such as transportation, shelter, and food, but their impact on the forest was minor. Not only were their populations relatively small, but most groups in this area were more focused on harvesting the ocean's rich resources, and not those from the woods, in order to survive.

Europeans, on the other hand, began clearing the area's forests when they started arriving on a permanent basis beginning in the early seventeenth century. Although the initial effects were very limited, significant development during the 1700s and 1800s brought a significant loss of forest cover. Moreover, Nova Scotia's distinct geographical attributes led to significant timber harvesting during these years. Its long, thin, rectangular shape rendered even its interior timber stands relatively accessible to the woodsman's axe, and also made it economical to deliver harvested wood to ports for shipment to the West Indies and overseas. In the process of selling its trees to the forest industry, the colonial – and then the provincial – government generally included in the sale the land on which the trees grew; other jurisdictions in Canada, by contrast, sold only the timber and retained ownership of the land. This created in Nova Scotia a patchwork of literally thousands of mostly very small timberland holders who had a strong incentive to harvest their trees. A badly degraded forest with a significant scattering of barrens (resulting from repeated forest fires or other natural conditions, such as shallow or water-logged soils) was the result.[3]

The situation again changed significantly with the coming of the pulp and paper industry just after the turn of the twentieth century. The industry's transition to using wood fibre instead of rags as its raw material suddenly imbued eastern Canada's spruce and balsam fir trees, even its smaller diameter ones, with a high value; Nova Scotians had been turning their larger diameter spruce and fir trees into lumber for decades, and a sizeable industry based upon exploiting them for this purpose had already developed. The availability of these species had also led to the establishment of several modest chemical and groundwood pulp mills in the province whose number and size were limited by several factors. These included Nova Scotia's lack – either perceived or real – of large, untapped water powers and large blocks of unalienated pulpwood. Finally, the province was located close to major American pulp and paper makers in the northeastern states and much of Nova Scotia's pulpwood was held in private hands and was thus exportable; most provinces had banned the export of pulpwood cut from "Crown," or government-owned, forests. The result saw a relatively large percentage of its annual cut of spruce and balsam – roughly 40 per cent during the 1910s and 1920s – being shipped to mills south of the border for processing.

There was one other significant factor about Nova Scotia's forests that made it stick out, literally, like a sore thumb.[4] Located on the northeastern tip of Cape Breton Island was the sole large, contiguous piece of unbroken Crown land, measuring over 520,000 acres (roughly 210,430 hectares) in size. It just happened to be covered in a mostly pure, luxurious blanket of fine pulpwood (mostly balsam fir). The provincial government was anxious to capitalize on its control over this valuable fibre supply – for good reason it was christened the "Big Lease" – to entice an entrepreneur into developing an industry to convert it into pulp and paper (Map 1). The Nova Scotia government signed an agreement in 1899 with a consortium of investors, one of whom was the aforementioned Frank J.D. Barnjum. The deal these investors made with the government was remarkably favourable to them, and it included permission to export pulpwood (as long as it was debarked locally), even if they did not fulfill their promise to build a mill. The businessmen intermittently cut and sold small volumes of timber from the Big Lease over the next decade and a half, and Barnjum retained control over it. In 1917, he sold the Big Lease to interests associated with the Oxford Paper Company. Oxford was one of the largest producers of fine paper in the United States, and it was centred in nearby Rumford, Maine. It had already begun acquiring timberlands in Canada from which it could export pulpwood to its operations in the US, and all observers realized at the time and since that the firm had acquired the Big Lease for the same purpose. Within a few years, it was exporting upwards of 30,000 cords annually from the tract.[5]

So the situation in Nova Scotia's woodlands was fairly straightforward in the mid-1920s, a time when there were still few concerns about bugs. Although the forest industry was an important component in the provincial economy, it was heavily weighted towards small-scale lumber production. Its pulp and paper sector was tiny compared to those that existed and were rapidly growing in neighbouring New Brunswick and nearby Quebec and Ontario; Nova Scotia was, after all, still awaiting the construction of its first newsprint plant. One of the major problems was the poor state of Nova Scotia's forests, a condition on which a landmark study in 1912 by the Commission of Conservation harped. Another report a dozen years later had expressed grave concerns about the lack of timber under Crown control (only about 5 per cent) and the rising volume of pulpwood that was being exported to the United States from the Big Lease. Ernest H. Finlayson, the author of that document and the dominion government's acting director of forestry, concluded his comments about Nova Scotia with an inauspicious remark. In terms of accounting for forest depletion by insects, Finlayson declared that "the mere fact that the province has not experienced an epidemic such as the budworm infestation of New Brunswick and Quebec, conduces to the popular belief that Nova Scotia has no serious forest insect problems. While there is perhaps no reasonable ground for thinking that an outbreak of this kind is impending," he added with uncanny prescience, "it is nevertheless a fact that insects do serious damage in the forests of that province, as they do in every province of the Dominion."[6]

Finlayson may have been a budworm whisperer. No sooner had his words been published than at least one prominent resident of Nova Scotia – the aforementioned Otto Schierbeck – was sounding the alarm about an epidemic that had recently been discovered in the province's far northeastern reaches. A Danish born and trained forester, Schierbeck had worked in Canada and Cuba briefly in the early 1900s before returning to his homeland to serve as Second Chief of Forests for the Danish government. He landed in Canada again in 1920 to begin a three-year stint as Chief Forester with the iconic Canadian firm Price Brothers Limited in the Saguenay region of Quebec. In this capacity, he studied many aspects of the company's operations, including the problems it was encountering with forest pests.[7]

Schierbeck had a golden opportunity to outline his findings in 1922 when Frank Barnjum decided to sponsor an essay contest. Its genesis is unclear, but the event would mark the beginning of a tight working relationship between the two. Barnjum had sold the Big Lease on Cape Breton Island back in 1917, but he had retained ownership of large areas of timberland in Nova Scotia and Maine and was a leading forest industrialist in both provinces. In the early 1920s he launched a sensationalist forest conservation campaign that included as a main plank in its platform a call to end the export of pulpwood logs from Canada; the effort was as sincere as it was self-serving.[8] In 1923, Schierbeck had become Barnjum's de facto private forester, and Barnjum was acting as Schierbeck's political backer in Schierbeck's campaign to become Nova Scotia's first Chief Forester. Although Barnjum argued that the deplorable state of the province's woodlands was ample evidence of the crying need for someone to assume this position, it was no coincidence that it was also an office from which Schierbeck would potentially be able to further Barnjum's designs to construct the province's first large newsprint mill.[9]

Barnjum's essay contest in 1922 was where that congruence of interests had begun. He had boldly offered a prize of $5,000 to anyone who could devise a plan to combat the trifecta of forest insects – the spruce budworm, the bark beetle, and the borer – that were ravaging the commercial woodlands of northeastern North America, and he provided some startling data to demonstrate the pressing need to destroy these pests. Barnjum estimated – without citing a source to support his claim – that bugs had destroyed an incredible 150 million cords of pulpwood in Quebec over merely the last decade, a volume of wood fibre that represented forty-five years' worth of supply for *all* the newsprint producers in North America! Of the hundreds of papers that Barnjum allegedly received, he judged Otto Schierbeck's *Treatise on the Spruce Bud Worm, Bark Beetle and Borer* the best. Significantly, in light of future events, Schierbeck insisted that he share the contest's purse with three entomologists – J.M. Swaine, F.C. Craighead, and J.D. Tothill – who worked for the dominion government, and who had not even entered the competition but were conducting intensive studies of the budworm at this time.[10]

Schierbeck's essay zeroed in on both the budworm as the largely ignored bête noire of eastern Canada's commercial forests and the multi-pronged strategy he had devised for slaying the beast. He based his ideas on the investigative work he had done on the budworm – among other insects – in Quebec while he had worked for Price Brothers. He recounted the truly unfathomable damage that the pest had done to eastern Canada's pulpwood supply, and he confidently predicted that further studies of the budworm would lead to a solution. As grounds for his optimism, he pointed to the success scientists had recently enjoyed in combatting pests and vermin, including the Americans' victory over mosquitoes during the construction of the Panama Canal and the ability of his Danish brethren to clear Copenhagen of rats by infecting them with a deadly pathogen. He also cited concrete examples of cases where parasites had been introduced to tackle a menace in woodlands. For Schierbeck, herein lay the greatest potential solution. He was convinced that the same approach would work for dealing with the spruce budworm. "I feel certain," Schierbeck declared authoritatively, "that the budworm can be combated by developing and culturing its parasites and infecting the moths and larvae. In Europe, this method has been tried, sometimes with good results, sometimes with minor results … when the most effective parasite is found, war should be declared against all pests."[11]

Schierbeck's treatise was so much more than a paper about forests insects. He aimed to realize a far larger political goal, one that aligned perfectly with Barnjum's own. Schierbeck's prize-winning essay was a call to arms for Canadians who had heretofore been ignorant of the need to adopt – immediately – proper forest management techniques to address all the threats that their woodlands faced. They had to take protective measures against fire, disease, and insects, he insisted, and it was high time that Canadians managed their trees for long-term sustainability instead of mining them in a wanton manner. In effect, he was calling for the entire suite of forestry measures for which professional foresters had been lobbying for roughly a generation. The challenge, Schierbeck despaired, was that Canadians, just like Europeans a century earlier, would not recognize the need for such measures on their own. Only a "forestry panic," as he put it, would compel the country's citizens to act.[12]

And the Danish ex-pat was only too ready to foment that panic in late 1925. Its spark would be the spruce budworm outbreak in eastern Nova Scotia, and Schierbeck's actions would be a poorly veiled attempt to realize Frank Barnjum's political and industrial aims. To further his agenda, Barnjum was not afraid to change his political allegiances. Having been a long-time Liberal, by the mid-1920s his dismay with the Grits' refusal to ban the export of pulpwood from Canada had driven him over to the Conservatives. To seek revenge, he ran successfully as the Tory candidate in the riding of Queens southwest of Halifax in the provincial election of 1925. The next year, however, he resigned from his post when the government refused to facilitate his plans to construct a 200-ton newsprint mill, the first project of this scope in the province. Nevertheless, after he secured a seat in the provincial legislature, he was much better situated to create the office of Chief Forester for Nova Scotia and hand the position over to Schierbeck, but there was still resistance within his party to him doing so.[13]

This is when Schierbeck stepped to the plate. He was more than ready, willing, and able to unleash a maelstrom to create the requisite public support to invest him with the title of Nova Scotia's Chief Forester, and the budworm menace would set it churning. Soon after the provincial election in June 1925, John C. Douglas, the new attorney general for Nova Scotia whose portfolio included administering the province's Department of Lands and Forests, allegedly asked Schierbeck to investigate the pulpwood situation on Cape Breton Island. This survey included reviewing the condition of the Big Lease, which the Oxford Paper Company controlled and from which the firm was exporting thousands of cords of pulpwood annually. Schierbeck ensured that his report was publicized widely in the local and national press and all the country's leading forest industry publications, and it appears he even leaked it to the media before formally presenting it to the provincial government.[14]

To be sure, there was no denying that Nova Scotia was suffering from a budworm infestation that had grown to sizeable proportions. In fact, Malcolm Swaine, the dominion government's forest entomologist, had been tracking the insect's activity in the province for roughly a year. His colleague who had investigated the situation, and another who would shortly do so, had reported that the budworm epidemic was probably a few years old and was centred on the western half of Cape Breton Island; a small outbreak had also been detected on the mainland across the Strait of Canso in Guysborough County. Although it was clear that some areas had been attacked intensely (and within them the balsam fir had died as a result), the dominion's Entomological Branch had concluded that generally the insect was infesting the entire area only superficially. In other words, this specific budworm outbreak need not be considered particularly devastating or worthy of extraordinary attention.[15]

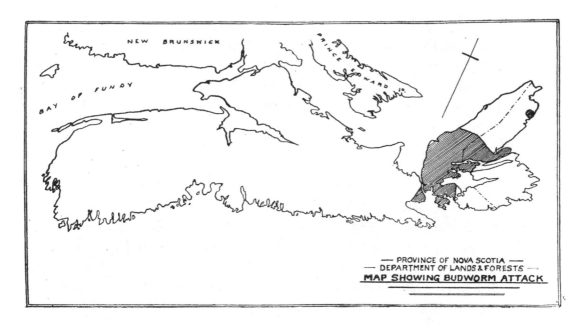

2.1 Map of Nova Scotia that illustrates the location of the spruce budworm Infestation, 1927.
Source: Nova Scotia, *Report of the Department of Lands and Forests, 1927* (Halifax: King's
Printer, 1928), 30A.

Schierbeck, of course, saw it very differently. He presented the Nova Scotia government with an incendiary report on the subject in the fall of 1925 that created the impression that the bug's presence signalled that Armageddon was at hand in the province's forests. His findings actually created headlines right across the country, with one in the *Globe* in mid-September screaming, "Budworm Scourge Menaces Entire Coniferous Forests." An article in the *Halifax Herald* entitled "Forests of Cape Breton Are Being Ruined by Plague" summarized Schierbeck's report. Redundantly describing him as a "forester and forest insect entomologist," it explained that he had found "the attack much more serious than he feared … A million and a half cords of pulpwood have been destroyed or are being destroyed through the ravages of this pest." Schierbeck declared that the outbreak had begun three years ago "and with no attention whatever has spread unchecked into alarming proportions."[16] Predictably, the coverage of the story provided by the *Sydney Record*, which was the leading local paper on Cape Breton, was even more extreme. In a front page article, it reprinted verbatim the final paragraph of the report Schierbeck had prepared for the government. "The present budworm attack," the newspaper quoted, "if not terminated, will wipe out the entire coniferous forest of the island thus jeopardizing the entire coal industry of Cape Breton and absolutely putting a stop to all future pulp and paper development, through the destruction of all pit prop and pulpwood material." Schierbeck's narrative stressed that the infestation was now within a few miles of "the only remaining virgin forest in Cape Breton," namely the Big Lease, and the provincial government's profound interest in this timber behooved it to take immediate action to deal with the situation.[17]

After having laid out in apocalyptic terms the extent of the impending calamity and highlighting how the authorities – presumably the now deposed Conservative provincial government and dominion entomologists – had negligently allowed the problem to fester since 1922, Schierbeck then outlined the few simple remedial measures he had devised to address the crisis. These measures echoed the recommendations he had made in his earlier essay on the budworm: introduce parasites to attack the insect and harvest the trees in which its larvae overwintered. Schierbeck's insinuation was unmistakable. The first step in implementing these measures was to appoint him Nova Scotia's first Chief Forester.[18]

Although this was a rare case of the theatre of maudlin forestry at its finest, its tragic flaw was that it lacked a defensible script. No sooner had Schierbeck broadcast news of the looming catastrophe than his views were being directly challenged. Swaine was out in Vernon, British Columbia, at the time and was shocked by an article he had read in the local daily. "I noticed in a western paper the other day," Swaine relayed to Schierbeck in late September 1925, "a statement from you to the effect that if the Bud Worm at Cape Breton continued it would mean the destruction of all the Balsam and Spruce on the island. Judging from the report that Mr. Gorham [a dominion forest entomologist] made last Fall I did not anticipate any such condition." Curious to know what Schierbeck had seen during his recent field investigation, Swaine explained that Gorham had reported only a few small outbreaks of the budworm and noticed only old budworm activity and not any active infestations. All the same, Swaine promised to dispatch one of his entomologists to Nova Scotia to examine the forest in question.[19]

In the meantime, Schierbeck's desire to land the job of Chief Forester for Nova Scotia drove him steadfastly forward. He was determined to appear to be taking clear-cut action to deal with the budworm scourge he had so sensationally promoted. To advance this goal, he outlined to John Douglas, the attorney general and head of Nova Scotia's Department of Lands and Forests, his immediate strategy. Just as forest fire fighters today cut fire breaks in advance of a blaze's progress in order to deprive it of fuel and thus contain it, Schierbeck recommended taking a very similar approach to the insect epidemic in the province's eastern forests. It would involve harvesting the affected areas on Cape Breton Island, comprising roughly 8,100 acres and measuring about 40,000 cords of pulpwood that were owned by a hodgepodge of small timberland owners and leaseholders. Not only would this operation salvage damaged and dying timber, he contended, but it would create a cordon sanitaire between the existing infestation and the rest of the forest, particularly the prized Big Lease; the budworm problem would thus be besieged and disappear.[20] Schierbeck also listed the names of the firms that would cut the timber and how much each of them would harvest. While he considered it prudent to decrease the fees they paid for the wood as an incentive for doing the work, going forward he called for the provincial government to introduce a new levy – it amounted to a "budworm protection tax" – on Nova Scotia's timber holders to pay for further control measures. Reasoning that the parties that leased or owned the largest swaths of forest would have the most to lose from future budworm outbreaks, he proposed a sliding tax scale that ranged from 1.5 cents to 4 cents per acre. For the Oxford Paper Company, which controlled the Big Lease, the bill would come to a whopping $24,000.[21]

While Schierbeck's forthright approach to addressing the budworm issue left no doubt that he was a man of action, there were several problems with his strategy for dealing with it. The most pressing one boiled down to a simple matter of dollars and cents. Initially, Oxford had been only too happy to cooperate with Schierbeck and the Nova Scotia government in dealing with the infestation that was threatening the Big Lease. Once the firms that controlled the timber designated for the salvage cut were informed of Schierbeck's plan, however, they balked at its cost. Not only was the market for balsam and spruce very poor at this time, but there was also little interest among the members of industry in having the government dictate how they conducted their business. The real fly in the ointment, however, was Schierbeck's radical new budworm protection tax. Immediately upon learning of it, the province's leading timber owners voiced their concerns to provincial politicians. The new timber levy was dead in the water.[22]

By this point, Oxford had decided it was time to flex its political muscle, and in this regard it was particularly well endowed. There was no denying that the firm had the requisite political connections in order to realize its aims. Its president, Hugh J. Chisholm Jr., was a longtime Republican, as was his well-known industrialist father. Republican Carroll L. Beedy also served as one of Maine's two elected officials in the US House of Representatives from 1921 to 1935, and he was the brother of Howard E. Beedy, Oxford's vice-president and treasurer.[23] Furthermore, Oxford had plugged itself into Nova Scotia's local network of éminences grises when it had purchased the Big Lease and thus become the province's largest timber owner by hiring the prominent Halifax lawyer James Ralston to represent the firm in Canada. Oxford exploited these contacts when it came back to the Nova Scotia government with a new plan to counter the one Schierbeck had already proposed.

Oxford's strategy involved seconding Frank Craighead. He was a leading American entomologist whom Swaine had hired away from the US government's Bureau of Entomology in 1920. While working for Canada's Entomological Branch, Craighead had teamed up with Swaine to conduct an intensive investigation of the spruce budworm and produce the first ever comprehensive report on the insect three years later. In 1923, Craighead had been lured back to the United States to become the Bureau's Chief Entomologist, but he remained closely connected to forest pest issues on both sides of the border. As a result, Oxford Paper had recommended in 1925 that Craighead travel to Cape Breton to inspect the budworm infestation there. This was a highly unorthodox suggestion, to say the least. The company was asking that an American scientist (specifically, a forest entomologist) employed by the US government conduct a study in Canada on behalf of an American firm operating in Canada even though the dominion government employed its own coterie of forest entomologists who were already involved in studying the matter in question! Although there is no paper trail to explain exactly how it happened,[24] Oxford was able to gain permission through the US Congress for Craighead to be given a "special leave" from the US Bureau of Entomology to head off to Cape Breton to undertake his reconnaissance.[25]

At first, Schierbeck was very supportive of Craighead carrying out this unique mission (he even convinced the Nova Scotia government to split the cost of Craighead's visit with Oxford Paper), but ultimately Craighead's report would not present news that Schierbeck wanted to hear. In early November 1925, Craighead travelled to Cape Breton Island with Schierbeck and C.W. Aeton, Oxford's local representative, to investigate the budworm outbreak, and Craighead produced a one-page précis of the situation by the middle of the month. It declared that "a serious budworm epidemic" existed in the southeastern part of Cape Breton, but that the situation farther north was "less serious. Some of the feeding occurred the past two years but all present evidence suggests that natural factors caused a high mortality in this years [sic] brood and that the insect may be even less numerous next season. In my opinion," Craighead stated matter-of-factly, "the situation does not demand the control measures suggested and that furthermore in consideration of the abundance of the budworm further south clear cutting of the strip proposed would not serve as an effective barrier for the country to the north." He closed by recommending that the situation demanded careful surveillance over the next summer.[26]

While the sting of this rebuke was still smarting, Schierbeck received another reprimand, this one from Swaine. First, as Canada's forest entomologist, Swaine was more than a little shocked to learn that the Nova Scotia government had been part of the effort to second America's Chief Entomologist to investigate the budworm situation in Nova Scotia. Fortunately for Schierbeck, Swaine took no umbrage from this series of events. Craighead was one of Swaine's closest professional colleagues and the latter had "every confidence," as Swaine put it, in the former's judgement. The problem for Schierbeck was that Swaine concurred wholeheartedly with Craighead's views and completely disapproved of Schierbeck's suggested remedies to the problem.[27]

Swaine outlined his thoughts in a lengthy, pointed letter to Schierbeck in mid-November. He first reiterated his promise to send one of his men down to Cape Breton to investigate the situation, and then hit upon the crux of the matter, bluntly expressing his regret that "it was not possible for me to confer with you before your recommendation for control was made." The reason was simple. Swaine believed that there was practically no chance that the "control cutting" would work and he had "never seen a case where I thought this method would be effective." Similarly, Swaine pulled no punches when he provided commentary on Schierbeck's recommendation to use parasites to combat the budworm. "I am sure it is useless to put any dependence on that method," Swaine put it pithily. "Tothill [another dominion forest entomologist] studied parasites of the spruce budworm in New Brunswick for several years and came to the conclusion that they played only a small part in the control of the spruce budworm outbreaks." To buttress his point, Swaine pointed out that the parasites of the budworm in Nova Scotia had had every opportunity to multiply recently, and the fact that the outbreak was occurring indicated that there were simply not enough parasites. He also underscored that the budworm was a native species, which meant that the only source of parasites would be the outbreak itself. "How, then," Swaine asked Schierbeck curtly, "would you improve upon nature's methods and where would you get your parasites?" Swaine added that another danger inherent in implementing Schierbeck's plan was ending up with hyperparasites, a result that could cause "as much harm as might be done by the transfer." Ultimately, Swaine declared that the "whole idea is absolutely impractical." Swaine added an afterthought that was significant in light of future events. After again reassuring Schierbeck that a dominion entomologist – M.B. Dunn – would be studying the situation in Nova Scotia as soon as possible, Swaine mentioned that one approach Dunn would carefully consider to deal with the problem was "the possibility of control by airplane dusting."[28]

Craighead and Swaine were acutely aware that their reports could potentially derail Schierbeck's plan of capitalizing on the hysteria he had created surrounding the budworm situation. In mid-November, Craighead sent Swaine the former's report on his recent visit to Cape Breton and added that, "no doubt by this time you know the circumstances governing my undertaking this detail." An exchange of letters between the two revealed that Craighead had gained a "somewhat different" impression of the politics of the situation than had Swaine, but that they were nonetheless in complete agreement about the mess that Schierbeck had created. "I am awfully sorry that Mr. Schierbeck pushed the proposition as he did," Craighead lamented to Swaine. "It will no doubt react quite unfavorably on his standing but of course it was absolutely impossible for me to agree with the proposals he had made. I will not discuss the question further at this time."[29]

Although Craighead had had enough of the "politics of this situation," Schierbeck certainly had not. He was incensed at the criticisms both Craighead and Swaine had levelled at his plans. Instead of recognizing the logical reasons for having done so and letting the controversy die down, however, Schierbeck decided instead to throw fuel on the fire. He informed John Douglas, Nova Scotia's Minister of Lands and Forests, that Craighead had corroborated Schierbeck's view that there was indeed a serious budworm outbreak on Cape Breton Island. In the same breath, however, Schierbeck indicated that the rest of Craighead's report was grossly unreliable and was tainted by a Yankee conspiracy. The problem, Schierbeck contended, was that Craighead had not proffered "an entirely unbiased opinion on this subject" because Oxford opposed the controlled cutting operation from the beginning "to minimize the danger so that all demands on their cooperation and purse might be avoided." Schierbeck pointed out that Oxford's president, Hugh Chisholm, had arranged to see Craighead in New York prior to Craighead travelling to Cape Breton and he alleged that, during their meeting, Chisholm had "managed to instil the idea in the Doctor [i.e., Craighead] that the budworm situation was greatly exaggerated and was utilized by the Provincial Government as a means to harass the company, this being the first step to force the company to abandon the big lease." Schierbeck went on: "I believe the company played on the 'fellow citizen' string, one American helping the other. I am positive from my conversation with Dr. Craighead … during our joint investigation that this, my idea, is right."[30]

Schierbeck then broadened his attack to smear Swaine and his colleagues as well. "The attitude of the Entomologist [Swaine] is incomprehensible," Schierbeck bemoaned, noting how Swaine and his team apparently threw up their hands in the face of an insect epidemic and collectively declared "that nothing can be done, in other words, 'let her rip.'" Schierbeck pointed out that the Americans had conquered yellow fever in building the Panama Canal, and that "we in Canada can devevlope [sic] sufficient energy to stop this pest, which, if not stopped, without doubt, will destroy all the coniferous forest of Eastern Canada. Through the machinations of the Oxford Paper Company," Schierbeck pronounced exasperatingly, "it is now impossible to undertake the control cutting around Baddeck." Although he explained that Swaine was keen to experiment with using insecticides dropped from planes to attack the budworm outbreak in Nova Scotia, he cautioned that "I have not much faith in the result."[31]

Schierbeck exacerbated the situation a short while later. He authored an article in the widely read *Pulp and Paper Magazine of Canada* that publicized to a national audience many of the criticisms of the dominion entomologists that he had already expressed to Douglas, and then some. His column ostensibly dealt with windfall as a disturbance factor in forestry, but Schierbeck used it as an opportunity to denigrate all entomologists – including those employed by the Canadian government – for being asleep at the switch, and to call – yet again – for proper forest management in Canada.[32] Veering specifically into the subject of the spruce budworm and how to stop its attacks, Schierbeck contended that "the entomologists seemed to have given up," and with indignation asked, "Are the Canadian people and forest industries going to let it go at that?" He stressed that the tragedy of the situation was that a budworm infestation could be arrested with one simple remedy. "The means are the parasites of the insect," he emphatically proclaimed. "The entomological department must wake up to its responsibilities. This Department is going backward instead of progressing. ... You will get the Government to wake up in this matter only if you demand action." The message was clear. If Nova Scotia wanted to prevent the budworm from wiping out its remaining stands of pulpwood, it had to take matters into its own hands by hiring Schierbeck to save the day.[33]

Soon enough he realized his long sought-after goal, a result that had a predictably settling effect upon his behaviour. Throughout late 1925 and early 1926, Schierbeck continued taking swipes at Ottawa for what he considered its negligence in addressing the few major forest insect problems with which Nova Scotia had been grappling over the previous decade.[34] He also convinced the province's Minister of Lands and Forests to write to the dominion entomologists to explain that Nova Scotia had decided of its own volition to cancel the control cut to contain the budworm menace on Cape Breton. It had done so *not* because it would be ineffective, the province's letter to Ottawa insisted, but because the winter's deep snow fall would not permit the control cut to proceed.[35] Then, in March 1926, at long last Schierbeck was appointed Nova Scotia's inaugural Chief Forester, and thereafter his manner stabilized considerably.[36]

Although Swaine responded to Schierbeck's calumnies, the former was too professional to get sidetracked by engaging in a public mud-slinging battle.[37] Swaine's focus remained strictly on furthering his field of study in general and addressing the budworm outbreak in Nova Scotia as much as his resources would allow. To calm relations with the Nova Scotia government, Swaine tried to set the record straight by communicating directly with John Douglas, who oversaw the province's Department of Lands and Forests. In doing so, Swaine described the budworm outbreak as decreasing in intensity and unlikely to spread northward into the Big Lease. He also explained in non-technical terms why, without mentioning Schierbeck by name, the idea of creating a control cut to contain the budworm epidemic was doomed to fail. Even clearing a ten-mile break in the forest around the infestation would be futile, Swaine pointed out, because "the spruce budworm moths will fly twenty miles or more very readily, provided suitable wind conditions prevail." He added that the only hope for controlling the outbreak lay in "the distribution of poisoned dusts from air machines," a technology with which he was eager to experiment in Cape Breton if he could procure the resources to do so.[38]

Swaine reinforced his views by sending Douglas a copy of the report that M.B. Dunn, a dominion entomologist, had recently prepared. Although Schierbeck had insisted that Dunn concurred with his own diagnosis of the budworm situation in Nova Scotia and his plan for treating it, Dunn's report demonstrated otherwise.[39] Like Swaine, Dunn had a lengthy history of studying the budworm and its activities in eastern Canada's pulpwood forests. Dunn and L.J. Simpson, another dominion colleague, had investigated the infestation for themselves and their conclusions about it were in accord. In their eyes, the outbreak was waning and would probably die out within a year or two, with Dunn further noting that "if it has not yet penetrated to the 'Big Lease,' is unlikely to do so." They also agreed that the proposed control cutting would not work and was of "questionable necessity" given their prediction that the attack would subside naturally.[40]

These events, particularly Schierbeck's crowning as Chief Forester, stilled the waters between him and all concerned parties, and thereafter it was remarkably smooth sailing towards dealing with Cape Breton's budworm infestation. In the early spring of 1926, Schierbeck – with the help of his political backer, Frank Barnjum – was able to draw unprecedented attention to the insect problem in particular and forestry in general. Together, they capitalized on "Save the Forest Week" (an American creation that Canada adopted in 1923) in mid-April to make issues affecting the province's woodlands front page news across Nova Scotia.[41] The provincial government exploited the hoopla surrounding the occasion to announce the creation of a new forest protection service, which was to be staffed by a team of trained rangers guided by Schierbeck's steady hand. He and Barnjum also managed to have cartoons that highlighted the many dangers that confronted Nova Scotia's forests figure prominently in the *Halifax Herald*'s coverage of the story. One caricature

2.2 Cartoon depicting the spruce budworm menace in Nova Scotia. Source: *Halifax Herald*, 20 April 1926.

(Image 2.2) was particularly poignant. It mocked those who jocularly dismissed the menace the budworm represented for being foolhardy politicos who, just like Nero fiddling while Rome burned, laughed while the pests gorged on the province's forests.[42]

Back in Ottawa, Swaine was doing all he could both to help Nova Scotia deal with its insect infestation and to raise awareness of the menace that the spruce budworm represented to eastern Canada's forests. In an article published in the *Forestry Chronicle* in 1926, Swaine provided the country's woodlands managers with a brief history of and update on the spruce budworm's activities. Swaine reiterated that humans would never be able to eradicate the pest and that it was "practically impossible" to control a budworm outbreak once it had become "epidemic over vast areas of forest." As a result, he declared that the only enduring solution was to alter harvesting strategies away from cutting spruce and instead to "utilize balsam [fir] as completely as possible." In the short term, he hoped that cutting or burning the infested timber might work to control the damage, but only for highly local-ized infestations, which were rare. He also felt the same way about aerial dusting, but there were specific problems associated with this potential solution as well. It would probably turn out to be very expensive, and it was potentially very danger-ous. "The risk to the pilot involved in flying low over forest land has thus far pre-vented adequate tests on a large scale," Swaine lamented, but he was optimistic that "a small, lighter than air, machine might solve the difficulty. Although this expense would be great," he reiterated, "a thorough test of forest dusting should be made as soon as possible."[43]

Swaine was also taking action to deal with the pressing spruce budworm issue in Nova Scotia specifically, and updated Schierbeck on his work there. He arranged to send a seasoned budworm specialist to the province during the spring and summer of 1926 to examine the situation and outline measures for addressing it. After reminding Schierbeck of the many reasons why the remedies Schierbeck had recommended – the control cut and releasing parasites – would not work, Swaine again relayed his hope that new technology might prove invaluable in cases where the budworm's population had erupted. "I believe that airplane dusting can be developed to be an efficient and practical method," Swaine explained, "but I have not been able yet to obtain tests on a large scale so that we could prove it. We came very near arranging for an extensive experiment last summer, we had even called for tenders for ten thousand tons of dust. Our plans were finally abandoned through the Air Board's decision that the work would be too risky." Ultimately, Swaine informed Schierbeck that there would be no silver bullet for the budworm, and that the only lasting solution would be to concentrate harvesting operations in pure stands of mature and over-mature balsam. "If the trained foresters and pulp-wood owners do not carry the matter further," Swaine asserted, "and find a way of keeping the old balsam utilized and if possible increase the percentage of spruce in the stands, the fault will lie with them. That feature of the problem is up to the forester and the pulpwood owner."[44]

But that was a macro perspective on a micro issue with which both Swaine and Schierbeck were dealing in mid-1926, and at that time they linked arms to move their effort forward on the ground. Aubrey H. MacAndrews, a dominion entomol-ogist who would soon earn his doctorate and take up a teaching position at the New York State College of Forestry in Syracuse, New York, spent much of the summer stationed in eastern Nova Scotia conducting his own study of the budworm.[45] On this mission he was accompanied for part of the time by Ernst J. Schreiner, Oxford Paper's research forester and future international expert on forest genetics and tree improvement.[46] In June, Swaine and Schierbeck also visited MacAndrews to get a first-hand look at the situation. The next month, senior officials from Oxford Paper cooperated with Swaine and Schierbeck to facilitate another visit by Craighead to re-examine the spruce budworm infestation; on this occasion there is no record of Schierbeck casting aspersions on the American entomologist's character. By this time, Swaine had submitted both his request to the Air Board to conduct, as Schierbeck put it, "experiments with dusting from balloon or aeroplanes" on Cape Breton and an application for a $10,000 grant to pay for the trial.[47]

The reports from Nova Scotia indicated that there was still a need to control the spruce budworm outbreak, but as its behaviour altered so too did the measures suggested for combatting it. MacAndrews reported that the infestation that had been threatening the Big Lease on Cape Breton Island had died back considerably, but the one in southern Cape Breton had spread westward across the Strait of Canso into Guysborough County on the mainland.[48] The season was far too advanced, however, even to consider an experimental dusting project. In addition, the turmoil created by the King–Byng Affair in national politics had also scuttled any hope of securing extra funding that year.[49] Attention thus turned to addressing other aspects of the problem. To deal with the aggressive infestation in Guysborough County, Schierbeck convinced Nova Scotia's government to push the area's largest timber owners to harvest as much of their infested balsam fir as possible in the hope of curtailing the insect's spread.[50]

At the same time, Swaine continued to pressure the Air Board to conduct the experimental dusting project in Nova Scotia and had good reason to grow increasingly confident that it just might work. The Americans continued to enjoy tremendous success in their aerial dusting projects over cotton fields in the Deep South, and more recently they had conducted significant experiments against pests that attacked trees. One such trial had been carried out in suburban Massachusetts, so it bore little relation to the work that the Canadians wanted to do over commercial woodlands. The other, however, could serve as inspiration for the project that Swaine was lobbying to conduct. Over 13–14 July 1926, officials in the United States oversaw the aerial dusting of 700 acres in the Peninsula State Forest in Wisconsin in an effort to combat the hemlock looper (*Lambdina fiscellaria*).[51] For reasons that are unclear, this project was almost immediately forgotten; the literature on this subject rarely even mentions it.[52] Moreover, the treated tract was certainly no natural forest. It was bisected by roads and broken up by many large clearings and homesteads that were still being farmed. It was also carried out for aesthetic and not industrial reasons. Nevertheless, it provided definitive proof that an aircraft had the capability of flying not far above forest trees and bombing forest pests with toxins. To learn more about this project in particular and the American aerial dusting work in general, Swaine headed off to the meeting of the American Association for the Advancement of Science in early 1927. His hope, as he told Schierbeck on the eve of his departure, was to use the knowledge he gained in the United States to make "a definite proposal" to the Air Board for an aerial dusting project in Nova Scotia when he met with its officials. "I hope we can arrange," Swaine explained, "to have a thorough test of that method. If it proves to be feasible to control budworm outbreaks by this method, even though the expense involved is fairly great, it will give us a weapon for use at least in special instances."[53] Around the same time that Swaine was expressing these views, word came from Europe that aircraft had recently been successfully employed to dust "several thousand hectares of forest" in Germany and France.[54]

Although Schierbeck had, only a few months earlier, obstinately defended his own plans for beating back the budworm in Nova Scotia and scoffed at the very different ideas presented by other entomologists, now Schierbeck was belting out songs from their hymn book. In late 1926, he asked Swaine for permission to publish in Nova Scotia's annual report for its Department of Lands and Forests one of Swaine's recent summaries of the budworm situation in the province. Swaine obliged, and the document appeared shortly thereafter. In it, Swaine expressly declared that the simple fact that the budworm moths could fly significant distances convinced him that no "programme of cutting" would be feasible to control the insect's destructive ways. He thus concluded that the best "and only logical policy, for both the Provincial Government and the Pulp and Paper Companies with permanent interests in Cape Breton," was to encourage the rapid harvesting of infested balsam fir and spruce. He also cast doubt on the potential effectiveness of using parasites to combat the budworm. Ultimately, Swaine expressed optimism that there was "a possibility of effective control through airplane dusting" and that he was hoping to test this approach the next season in Cape Breton. In retrospect, it was hard to believe that Schierbeck could countenance, let alone ask to publicize, these views that he had so recently dismissed as heresy.[55]

But even the synergy generated by this meeting of the minds could not convince the Air Board that it should undertake the experimental dusting project. The campaign to realize this end still needed an infusion of political strong-arming to push it over the top. It came in spades very early in 1927, almost entirely because of one politician's profound political and social ties to Nova Scotia. James L. Ralston was a native of Amherst, a manufacturing centre located on the isthmus that joins the province to New Brunswick. After graduating from law school in 1903, Ralston established his practice and continued it even after entering provincial politics. He distinguished himself while serving during the First World War, and rose to the rank of colonel thereafter. While practising law during the early 1920s back in Halifax, Ralston acted on behalf of the Oxford Paper Company and other American firms during the hearings organized by Canada's Royal Commission on Pulpwood. In this capacity, Ralston had aggressively defended his clients' right to export spruce from Nova Scotia, and he continued to advocate for them during the mid-1920s.[56] On the eve of winning by acclamation a by-election for the riding of Shelburne–Yarmouth in November 1926, Prime Minister William Lyon Mackenzie King appointed Ralston Canada's Minister of Defence. Holding this cabinet post meant that Ralston exercised authority over the Air Board, the leaders of which had repeatedly stymied Swaine's plans to catapult Canada's war on bugs into the modern era.[57]

Ralston would soon change all of that. In early January 1927, Howard Beedy, who managed the Big Lease on Cape Breton for Oxford Paper, conferred with Swaine about the threat the budworm posed to the firm's timber holdings in Nova Scotia. After their meeting, Beedy wrote Ralston to apprise him of the situation. Beedy explained that Oxford Paper would naturally continue to monitor the budworm threat and that the "only recommendation as to a method of eliminating the bud worm and preventing its spread, which seems practical is the one now being advocated by Dr. Swaine. He proposes to dust the trees with a chemical powder by means of aeroplanes and feels that this will kill the bud worm." Beedy added that this approach had been used successfully in the United States to kill "cotton parasites," and informed Ralston that Swaine was returning to Ottawa the next day to attend a meeting where he hoped to be "successful in securing the cooperation of the Air Board in this work." Beedy then hit upon the nub of the matter. He and Hugh J. Chisholm, Oxford Paper's president, "felt that it would be advisable to communicate with you [i.e., Ralston] and keep you informed, as Dr. Swaine may find it necessary to go to you and attempt to enlist your support. If you find it possible to assist Dr. Swaine in any way, I am sure you know that your action will be very much appreciated by us. You will understand, of course, that we are very much interested in seeing this bud worm epidemic stamped out as it exists in property adjacent to our holdings and constitutes a menace to our property."[58]

Although Ralston had never heard of the spruce budworm issue to which Beedy referred, the Minister instantly committed to doing everything he could to facilitate Oxford Paper realizing its aims. "I will be in Ottawa next week," Ralston reassured Beedy in a letter in mid-January, "and I need hardly say I will be glad to assist in any way I can. I expect that when I get to Ottawa I will hear of some application from the Department of the Interior to my Department respecting our co-operation in connection with the air services."[59]

Thereafter, Ralston proceeded full speed ahead. He contacted officials from the RCAF to learn all that he could about the matter, and asked if the Air Force could take on the work and pay for it out of its budget for the current year.[60] Next, he saw to it that Swaine prepared another request for the Air Board's help, one that focused solely on the situation in Nova Scotia. It reiterated the case Canada's forest entomologists had long been making, namely the pressing need to undertake a proper trial of aerial dusting against the budworm to see if it would work as a method of direct control. Swaine asked specifically for the Air Board to hire an outfit, preferably one that had honed its dusting skills flying missions over Dixie's cotton fields, to conduct a 500-acre experiment in Cape Breton in early June 1927.[61] The Department of Agriculture would underwrite the cost of the dust (roughly $1,500) needed for the effort, and Nova Scotia's Department of Lands and Forests and the Dominion Forest Branch would cooperate with Swaine and his crew in planning and conducting the operation. Explaining that Canadian entomologists had devised methods of forest management that would enable the forest industry to avoid major losses to the budworm in the future, Swaine emphasized that there was an "urgent need" for a direct method of control. "Airplane dusting," he declared, "offers the only hope." Swaine's memorandum was quickly passed up the chain of command, landing ultimately on Ralston's desk.[62]

Although the forest entomologists had been delivering nearly identical requests for help to the Air Board for the past several years, this time Ralston ensured that the entreaty was favourably received. In far less time than it takes for a spruce budworm to devour a fir needle, the Air Board's attitude swung from opponent to advocate of the proposal. Howard Beedy, Oxford's senior local official, assured Swaine in early 1927 that the Air Board now supported the dusting proposal and, not surprisingly, so too would "the Minister [i.e., Ralston] when the same is presented. I am sure," Beedy postulated, "that Mr. Ralston's earlier contact with this property, and its problems will put him in a very favourable frame of mind." Although the precise meaning of Beedy's cryptic message is unclear, in short order Ralston had leaned on the DND's Deputy Minister to find the money within the RCAF's budget to pay for the dusting work in Nova Scotia that season.[63]

Politics of a different sort then entered the fray. As much as Ralston viewed his support for the project as a means of endearing himself to some of Nova Scotia's most powerful industrial clients, Canada's senior commanding officer also saw it as a way to captivate the country's attention on the potential of the flying services he could provide to the nation. Senior officials within the air force quickly realized that a sum of $25,000 was available for the RCAF to hire contracting firms to carry out projects that it either would not or could not undertake. James H. MacBrien,

Canada's Chief of Staff, had other ideas, however. He regarded it as a golden opportunity both to purchase a special aircraft for dusting work and to begin conducting
this type of operation. In his view, the forest entomology project in Nova Scotia
would undoubtedly lead to others "requesting similar work in other areas of the
country." In the same breath he emphasized that "useful work of this kind will mate-
rially increase the support by the Public of the various activities of the R.C.A.F."[64]

When James S. Scott, the RCAF's director, presented sound reasons for not pur-
suing this course of action, MacBrien played the nationalist card to win the day.
Scott pointed out that the work could be done much cheaper by tendering it out,
and this would avoid the RCAF undertaking a serious risk; if the experiment failed,
the air force would be stuck with "an aircraft of doubtful utility for other work, and
of a type not standard in our Service." MacBrien would have none of it, however,
and he insisted that the RCAF should uphold its principle of conducting all its own
experimental projects. Scott continued to protest but eventually relented when it
became clear that no Canadian firms were in the business of aerial dusting.[65] "I am
not," MacBrien adamantly declared to Scott, "in favour of a contract being let to a
U.S. firm." Scott thus caved, and the RCAF committed to purchasing at least one
aircraft specifically for the dusting work.[66]

At this point, the natural environment demonstrated that it, too, would have a
say in this affair, specifically in terms of increasing the danger quotient for Canada's
inaugural forest dusting missions. Senior officials within the DND contacted Bert
Coad, the American entomologist in charge of the cotton dusting experiments in
Tallulah, Louisiana, in early March 1927 to ask if Coad would help train the pilots
who would undertake projects that season in Nova Scotia and Manitoba (the latter effort was going to target wheat rust). Coad greeted with alacrity the Canadians' request to train their pilots. In mid-April he advised that one week "should be
ample time provided we have a reasonable amount of flying weather."[67] The problem
was that that spring saw exceptionally heavy rains soak the Deep South, causing
the Mississippi River and nearby waterways to overflow their banks. The weather
and ensuing chaos were so severe that the dominion's two pilots, Flying Officers
T.M. Shields and C. Bath, were schooled almost entirely while on the ground "in
the theory of dusting," and experienced practically no flying time. Coad was thus
unconvinced that they were prepared to undertake the work in Canada, a perspective that he conveyed in a letter to the dominion's senior air force officials:

Unfortunately both of these men have apparently been flying other types of equipment
for a considerable period and neither of them are now at all familiar with flying the
highly maneuverable ships required for dusting. We gave them as much dual control

instruction along this line as we could but our chief pilot, who is a man with seven years of experience as an instructor in our Army training school, reports that he does not consider either one of these men safe to attempt this class of flying without further training and practice. Mr. Shields is considerably superior to Mr. Bath in this flying but even he needs some additional training and experience before attempting dusting.[68]

With time getting short – the dusting of the budworm outbreak in Nova Scotia was scheduled to begin in early June – the resulting pressure combined with a number of other factors to create what seemed like a never-ending comedy of errors. Coad wrote directly to Swaine in late April to relay his serious concerns about the Canadian airmen, Flying Officers Shields and Bath. After praising them for their "earnest desire to learn," Coad hit upon the crux of the matter. "Frankly though," he conceded, "I am much disappointed in their flying and really feel that if either of them attempted to start out with a dusting ship now, he would crack it up in short order."[69] By this time, Coad's earlier ominous assessment of the two pilots' aptitude for dusting – or lack thereof – had spurred senior officials within the RCAF to take corrective measures. They had ordered both men to rush down to the Keystone Puffer's production facility in Bristol, Pennsylvania, to pick up the two planes that the Canadians had ordered. Although they were offered flying lessons before they left the aerodrome, it hardly solved the problem. When senior government bureaucrats heard that the instructional sessions would cost a whopping $15 per hour, they told the Keystone officials that the aerial training was "not to exceed three hours."[70] Moreover, the Canadians had assumed that the Puffer aircraft had dual controls, whereby the instructor would be able to fly with his student and deliver hands-on training, but this had been a faulty assumption. "The machine does not have dual controls," Flying Officer Bath reported to headquarters. "The controls and instruments are merely interchangeable from one cockpit to the other." As a result, Bath had resolved to fly the machine from Pennsylvania through New England and up to Nova Scotia without *any* instruction and without wasting precious time, simply informing his superior officers of his amended itinerary.[71]

Despite these setbacks, there was no stopping the dusting project in Nova Scotia from proceeding, and during the spring of 1927 a veritable tidal wave of interest in using chemicals to kill insects was sweeping over the province. Schierbeck was now portraying himself as the true authority over and driving force behind the plan to coat the spruce budworm with deadly toxins. "Spraying the forests of Nova Scotia from aeroplanes, in an effort to destroy destructive parasites [i.e., forest pests], is a new idea that will be put into practice next month by Otto Schierbeck, Chief Forester," boasted the *Halifax Herald* in late May. While the article mentioned that

Swaine would be present to help organize the project, it emphasized that the work would be "personally supervised by Chief Forester Schierbeck."[72] At the same time, Nova Scotia's largest daily was replete with articles that exhorted both home gardeners and commercial farmers in Nova Scotia to kill their unwanted pests with chemical concoctions. "Spray and Dust Must be Kept Up," screamed one self-help headline. Another article described a plan to form "Spray Circles" in an effort to rid the province's fruit orchards of destructive pests.[73]

The *Herald* and other major dailies across the country also added to the excitement surrounding the imminent use of chemicals against Nova Scotia's forest insects by treating a few of the principals involved in the enterprise as celebrities. For example, they provided detailed coverage of the trip that Flying Officer Bath and his mechanic made in mid-June to pick up the Keystone Puffer down in Pennsylvania and fly it back to the province. The aircraft was to be used, a reporter for the *Herald* explained, "to aid in forest conservation work by scattering poison from the air in the war against tree destroying insect life." And in response to the reporter's question about the project's potential effectiveness, Bath stated that aerial dusting had proven "highly satisfactory" when conducted over the American Cotton Belt, and that "he expects the same will apply to the forests of Cape Breton."[74]

Heretofore, the only operation that the dominion forest entomologists had agreed to conduct was on Cape Breton Island,[75] and it was carried out a few days after the summer solstice. The community of Whycocomagh, on the west side of Bras d'Or Lake and one of the few local stops on the railway, served as the base for the operation. Over the course of roughly a fortnight, the Puffer dropped nearly three tons of poison – mostly calcium arsenate but also a small amount of lead arsenate – over Crown and freehold land just north of Orangedale, another nearby town (see Map 1). The target was a series of relatively small plots that the entomologists had laid out, the largest measuring 800 feet wide by 2400 feet long in total and divided into six equal-sized tracts. The aim was to deliver a different dose and type of poison dust to each plot, all in an effort to determine the most cost-effective means of killing the budworm.[76]

From the outset, the goal of this inaugural experiment had been strictly to learn how to conduct aerial dusting operations over natural forests. In almost every respect, the Canadians were starting from ground zero. Only a few of the lessons that Coad had learned from his dusting experiments in the Deep South were applicable to the work they were doing in Nova Scotia, and virtually no one even mentioned the project that had been undertaken in Wisconsin in 1926 nor any lessons that might have been drawn from it. As the dominion entomologists described

2.3 Airplane dusting on Cape Breton Island. Source: Nova Scotia, *Report of the Department of Lands and Forests, 1927* (Halifax: King's Printer, 1928), Figure 2.

their pioneering effort in Nova Scotia, "the results sought in this initial experiment were to solve the mechanical difficulties entailed in placing dust accurately and in different degrees of concentration on the areas selected and, through applying different doses of poison on different plots, to study the effect of the poison in killing the caterpillars."[77]

Reconstructing the practical steps involved in carrying out the venture provides a glowing testament to how the Canadians were bona fide innovators in this field. They determined, for example, the rate at which the dust was delivered from the aircraft by conducting a series of experimental flights using a definite number of pounds of the chemical and timing how long it took to empty the hopper (the compartment on the plane that stored the dust before it was released over the drop zone). The Canadians also learned that the hopper would not dispense the dust evenly if it were less than one-third full, and so this issue became a limiting factor in using it. Furthermore, Swaine described how they realized almost immediately that it was "useless to apply the dust unless the air was practically still. Even a very light wind would carry the dust cloud away from the swath where it should light or even entirely [drift] off the plot." Although there was often a calm period in the late afternoon, Swaine described how the flying conditions at that time were "much less desirable on account of uneven air pressure and it was believed better not to rish [*sic*] flying so low to the trees at that time." The lesson was clear. "The early morning

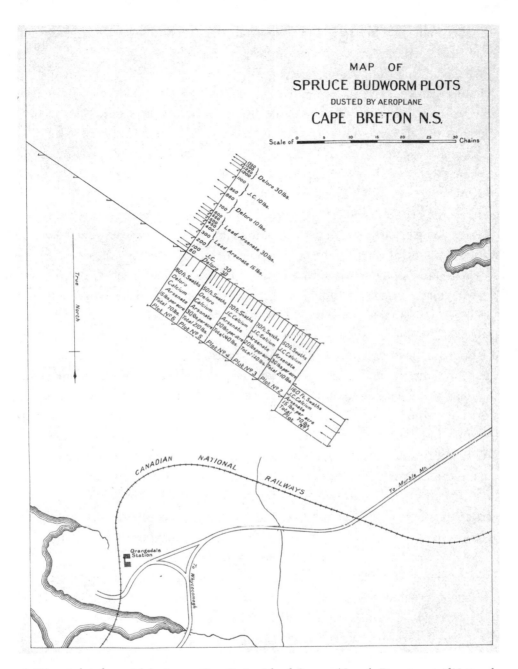

2.4 Target plots for aerial dusting on Cape Breton Island. Source: (Canada Department of National Defence, *Report on Civil Aviation and Civil Government Air Operations for the Year 1928* (Ottawa: F.A. Acland, Printer to the King's Most Excellent Majesty, 1929).

hours immediately after daylight and while the dew was still heavy on the foliage," Swaine declared categorically, "was the proper time for dusting."[78]

An equally high hurdle to be overcome was developing a system for demarcating the plots that would be dusted so that the entomologists could learn how different doses of poison affected the target pest (the goal was to determine the smallest, and thus the cheapest, amount of dust needed to kill the budworm). While the crews wisely cleared strips of forest to indicate the boundaries of each tract, the clearings were practically invisible to the pilot until he was directly above them because he would ideally fly only twenty feet above the crown of the forest. To address this issue, the entomologists realized they had to place markers above the treetops using some sort of material that the pilot could easily see from far away. It was hard enough for one of the entomologists' agile assistants to scale to the very crown of some of the tallest spruce and balsam trees and affix an object to them, but what made this mission doubly difficult was figuring out which type of material would work best under the circumstances. "At first we employed strips of factory cotton 8 feet long and 1½ feet wide wrapped about the top of the trees selected for the purpose," Swaine reported a few months after the mission was completed. "It was found," he continued, "that while the pilot could see these readily from a con-siderable altitude they were not atall [sic] easily distinguishable when he was flying close above the tree tops, particularly if they had been blown about and beaten in the wind and rain. It was found necessary to mark the plots by placing streamers on poles projecting at least 10 feet above the tops of trees." Because the pilot could only dust in calm weather, Swaine pointed out that these streamers would, at that time, be "perfectly vertical," but that "they are nevertheless visible from above and in the vertical position are more easily distinguishable by the pilot when flying at great speed immediately above the tops of the trees. The vertical position of the markers is more serviceable to the pilot than a horizontal position."[79]

In addition, there was one very surprising aspect to this whole enterprise. Although the entomologists had learned a great deal about how to conduct this type of project, and they had completed their initial effort, the aerial campaign against the spruce budworm in Nova Scotia was not done; politics would once again shape how it proceeded. The dust had barely settled on the areas that had been treated at the behest of the Oxford Paper Company when another major American paper firm with large timber holdings in eastern Nova Scotia prevailed upon the provin-cial government to dust the infestation that was threatening its wood supply. After all, Schierbeck and the dominion entomologists had been tracking the budworm's advance from Cape Breton across the Strait of Canso towards the mainland. Spe-cifically, the insect had established several beachheads in Guysborough County,

2.5 Man up in tree to guide airplane. Source: Nova Scotia, *Report of the Department of Lands and Forests, 1927* (Halifax: King's Printer, 1928), Figure 1.

2.6 Pole with streamer on top of tree to mark dusting plots. How the ground crew attached the pole and streamer to the top of the tree is unknown, but it was truly an extraordinary feat. Source: LAC, RG24-1-a, 4912, DND – Operations for Department of Agriculture, vol. 2.

which had drawn the attention of the West Virginia Pulp and Paper Company, or Westvaco.[80] In the early 1920s, the firm had been seeking a new pulpwood supply for its flagship mill in Covington, Virginia. It had chosen to source its fibre from eastern Nova Scotia, and had acquired tens of thousands of acres of woodlands in the area. Westvaco harvested and shipped the timber using its subsidiary, the Sonora Timber Company, which was named for the town at the centre of its operations in the province (Map 1 and Image 2.1).[81]

Westvaco had been politically astute in setting up its operations in Nova Scotia, a strategy that would pay off handsomely in tackling the budworm. It had engaged one of the Maritimes' most powerful law firms – Henry, Rogers, Harris and Stewart – which was based in Halifax, to act on its behalf in the province. Moreover,

Westvaco had firmly allied itself with the Oxford Paper Company. During the interwar years, the two companies occupied the same office buildings in Manhattan, first on Fifth Avenue and then later on Park Avenue, and the executives from both companies kept in close contact.[82]

Westvaco was thus able to use its political and economic might to leverage a favour from the Nova Scotia government and the dominion Air Board in what arguably became a quid pro quo arrangement. In the fall of 1926, Schierbeck had directed the company to begin harvesting the budworm-infested pulpwood in Guysborough County as one means of controlling the local infestation. In return, it appears that the Nova Scotia government agreed to arrange and pay for a significant dusting project to be carried out on Westvaco's local timber limits.[83] As a result, immediately after Flying Officer Bath had completed his mission on Cape Breton Island in late June 1927, he was directed to take off again to repeat the exercise a few dozen miles to the southeast using Mulgrave, on the Nova Scotia mainland, as his base. His superiors also ordered him to carry out "any reasonable request" that Swaine made of him "for additional work," even though Swaine took no real interest in the effort.[84]

Although this second project proceeded virtually under the radar, it was exponentially larger than the first one. Whereas the dusting project on Cape Breton had been a precision operation using exact measurements of poison and dusting plots, the one in Guysborough was akin to a carpet-bombing operation. Schierbeck saw to it that the provincial government provided twelve tons of dust to be used to cover blocks that measured one-half mile wide by two miles long and that "surround[ed] the whole infested area."[85] The operation lasted roughly three weeks and was constantly hindered by unfavourable weather conditions. As Schierbeck boasted to a concerned citizen in July 1927, "the dusting in Whycocomagh [i.e., Cape Breton] [wa]s purely experimental," whereas the project in Guysborough "was on a commercial scale."[86] The Nova Scotia government paid for the dust at a cost of $2,000 and the Air Board provided the aircraft and flying personnel; there is no record of Westvaco having contributed any funds to the cause.[87]

Initially, the giddiness from having conducted these groundbreaking projects tinted the glasses of its principals in a thick, rose-coloured hue. For his part, Swaine felt as though he had been part of a game-changing effort. Soon after the mission in Guysborough had wrapped up, he wrote to the RCAF's Controller of Civil Aviation, John Wilson, to express his profound gratitude for the assistance that the air force had provided on the experiment and to offer his views on its outcomes. Swaine was ebullient in informing Wilson that they had successfully "learned how to carry out such an experiment," and that the results so far "lead us to hope that we may be able to develop an effective means of control through airplane dusting." Swaine continued proudly, "We can say definitely that we succeeded in applying the dust successfully and that we have obtained a marked degree of success in killing the caterpillars." He also heaped praise on Flying Officer Bath for his adept handling of the assignment. "He took a very great interest in the problem we had to solve and did everything he possibly could to further the work. … He proved to be a marvellously skilful pilot and flew the plane on the dusting flights with remarkable accuracy."[88] Swaine delivered a similarly sanguine message in an article he published in the *Forestry Chronicle* in early 1928. "The possibility of success appeared to be sufficiently great to warrant a series of thorough experiments," Swaine asserted. "If a method of dusting could be perfected by which the budworm could be destroyed effectively, even though at considerable cost, it might prove to be of incalculable value in protecting valuable stands of timber or even in checking the development of incipient outbreaks."[89]

In many respects, Swaine had good reason to be so buoyant. His inaugural forest dusting trial garnered international attention. Even before it had begun, the British Air Ministry had caught wind of it and seemed perturbed by the colony's apparent impudence in taking the lead in this field without informing Westminster of Canada's intentions. Once Swaine had completed the undertaking and the *London Times* had published an article about it, British air force officials were downright anxious to learn "every detail possible of experiments that have been carried out." Moreover, the British wished to bask in the reflected glory of their colony's accomplishments. As one of their senior officers put it, the matter was "considered … of great interest to this country, and in fact, to the whole Empire."[90] Likewise, American forest entomologists – some with the United States Department of Agriculture – were still asking their northern neighbours for information about the dusting effort in Nova Scotia long after its completion.[91]

Those more distant from the projects, and in particular those who had something to gain from highlighting their success, were even more inclined to provide reports that glowed white-hot with optimism. The Department of the Interior, for example, was determined to broadcast to the world the cutting-edge work Canadians were doing in applying the latest aviation technology to solving problems in their commercial woodlands. William W. Cory, the Deputy Minister, was one of Canada's representatives at the Third British Empire Forestry Conference in Australia in 1928. In preparation for his trip, he encouraged his counterpart at the Department of National Defence to provide him with a film that he could show at the conference to help portray Canada's forestry program "in as favourable a light as possible." In that same letter, Cory bragged, "Canada leads the world in the use of aircraft for forest protection purposes [i.e., forest fire and entomology work]."[92] John Wilson of the RCAF echoed Cory's perspective. In describing the experiment to a British colleague, he emphasized that Swaine had been the driving force behind it and that the "preliminary reports show that the dusting was carried out successfully and that thousands of caterpillars have been killed." Although Wilson added a cautionary note – that it was "too early yet to state definitely whether they have been completely exterminated and the defoliation entirely stopped" – British officials demonstrated a discernable lack of interest in hearing about such qualifiers.[93] Finally, the *Pulp and Paper Magazine of Canada* broadcast enthusiastically that the "War on Forest Pests [had] Started." Even before the results from the effort had been tabulated, the piece affirmatively declared that "a sweeping victory over the insect pests which infest Canada's forests doing many millions of dollars['] worth of damage annually is expected as a result of recent experiments in dusting the woods with poison spread from a specially designed airplane."[94]

To be sure, there was much to applaud in terms of the dominion's initial dusting experiment. Leaving aside the monumental accomplishment of having overcome myriad practical challenges, one major reason to celebrate was the unprecedented level of cooperation that had occurred among a range of agencies and interested parties concerned with forest management in Canada. The dusting campaign in Nova Scotia had witnessed several arms of the dominion government – the Department of Agriculture's Entomological Branch, the Department of National Defence's Royal Canadian Air Force, and the Department of the Interior's Dominion Forest Branch – work with Nova Scotia's Forest Branch and officials from the Oxford Paper Company to carry out the first trial on Cape Breton; most of these partners had also been involved in the second one carried out on behalf of Westvaco. With this much synergy created, there was a sense that the financial burden for future operations could be shared broadly and thus their feasibility increased.

After passing through this honeymoon period, however, Swaine and his colleagues slowly began recognizing that it was time for a sober second thought when it came to evaluating the effectiveness of the dusting projects in Nova Scotia. First, there was the simple matter of dollars and cents to consider. "The cost of the operation," Swaine stated laconically, "might be prohibitive."[95] Moreover, the initial intent of the experiment had been to establish the lowest possible concentration of poison needed to kill the spruce budworm, and figuring this out depended upon being able to tally how many caterpillars each dose of toxic dust killed. But herein lay the problem. In a typical infestation, the insects literally covered the tips of the trees' upper branches; the plan had thus been to climb up and count the bugs before and after the poison was dropped. Swaine and his colleagues soon realized that this was a hopeless approach, however. "I have no further faith in checking results by caterpillar counts," Swaine despaired to Schierbeck after the initial effort on Cape Breton was over. Swaine had been forced to leave the project in mid-stream, but he described how, "by the time I left a man could not get half way up a balsam tree without dislodging nearly every caterpillar it contained. When we commenced that work we could actually count the caterpillars on the foliage; but after the dusting was done they were mostly large enough to drop when dislodged and I shall not place much value at all on the counts made on those trees after dusting." As a result, the scientists had to resort to rather unscientific methods to determine their body counts. "There is only one way to judge the results in this case, so far as I can judge now, and that is by the defoliation that takes place [next spring]," Swaine conceded.[96] Likewise, Aubrey A. MacAndrews, who had arguably been the entomologist most closely involved in the spruce budworm project in Nova Scotia, submitted a preliminary report on the dusting in January 1928 and he echoed Swaine's earlier comments. In paraphrasing MacAndrews's words, Swaine informed Schierbeck that "the work on Cape Breton Island shows less [sic] differences in results as between the different plots than we had hoped to obtain ... but we succeeded in learning how to conduct such an experiment, at least."[97]

By early 1928, Swaine was beginning to realize that the success of the dusting project in Nova Scotia had been undermined by different elements, largely of the non-human world. In his official report on the experiment, Swaine recounted the history of spruce budworm outbreaks in eastern Canada. In doing so, he repeated his earlier calls for forest companies to make it a priority to harvest mature and over-mature balsam trees as a steadfast means of combatting infestations. If they followed his advice, he argued, it would "eventually produce a condition of permanent immunity to spruce budworm outbreaks." In the meantime, he believed that aerial dusting offered the possibility of controlling outbreaks in "the more valuable areas," but he underscored that there were several major hurdles that still had to be overcome if this approach was going to work; these hurdles involved both the life cycle of the spruce budworm and the dynamics of spruce–fir forests.[98]

In terms of the first factor, Swaine and his team already agreed that it was best to apply the poison dust when the caterpillars were small because that was when it was easiest to kill them. The problem, however, was that these pests feed by burrowing into a tree's buds before the buds have fully opened, and "at that time [the budworms] are ... in great measure protected from the dust." Herein lay a second obstacle presented by nature. Although balsam fir often occurs in pure stands, it is frequently mixed with other conifers such as red spruce, white spruce, and black spruce, all of which open their buds at different times. "When the buds of balsam fir are open sufficiently to permit the dust to penetrate to the axis," Swaine pointed out, "those of red spruce, which open from one to three weeks later, are usually closed or only partly open, so that it will often be difficult to dust a mixed stand effectively." Furthermore, there was only a short period "in the early season" when the spruce budworm was most vulnerable to the poison, but Swaine and his cohorts had learned this was the time when "weather conditions are likely to be unfavourable."[99]

It was also now clear that another of the budworm's foes had begun taking a toll on the infestation in Nova Scotia's eastern forests, just as Swaine and his colleagues had predicted. "Within a few days following the dusting," Swaine reported, "the caterpillars on the plots were found to be dying in considerable numbers, and this mortality continued for the following two weeks." While these results arguably spoke to the effectiveness of the poison dust on those plots that had been targeted, the same condition was apparent, "though to a less marked degree," on the "check plots," which had *not* been treated. "It soon became evident that a few of the caterpillars were killed by fungi and that a very large number of them were parasitized," Swaine admitted. "A subsequent study showed that on the whole area infested by the budworm, the parasitism reached as high as 75%. It was, therefore," he regretfully concluded, "exceedingly difficult to determine the precise effect on the caterpillars to be ascribed to the poison."[100]

Schierbeck delivered this same news to J.A. Duchastel, manager of the Quebec Forest Industries Association, who was most curious to learn whether the dusting operation had worked. "Last summer the budworm was very strongly attacked by parasites, in places as much as ninety percent of the larvae and cocoons showing such attacks," Schierbeck explained. "These attacks have, especially in Cape Breton, practically annihilated the budworm and strongly decimated them in Guysboro County. It is, therefore, very difficult to say what effect the dusting had on the budworm, since the deadening due to the parasites was so overwhelmingly great." In his reply, Duchastel expressed his dismay that nature had beaten the entomologists to the punch in terms of combatting the budworm. "It is perhaps unfortunate from the point of view of the science of forest protection," he ruefully declared, "that this should have been chiefly due to natural causes, but I must congratulate you nevertheless on your escape from the danger of a really drastic devastation."[101]

Ultimately, Swaine realized that there was but a single lesson to draw from the entomologists' innovative work in Nova Scotia. As he cogently put it, he and his colleagues had "learned how to conduct such an experiment." Furthermore, Swaine was still convinced that this work was well worth the effort because of the potential panacea aerial dusting represented in certain entomological crises. "Next season," he confidently told Schierbeck, "we shall be able, I hope, to obtain definite results."[102]

But all these analyses had ignored the elephant in the room, a creature whose very size would soon make it impossible to discount. For as long as the aerial dusting projects had been discussed among the entomologists and officials from the Air Board, the latter had been profoundly concerned about the dangers inherent in dropping chemical dusts over forests. The inaugural flights over the trees in Nova Scotia had convincingly demonstrated that their fears were well founded. The pilot who had been at the controls – Flying Officer Bath – would soon relay his views to his superiors, and the conclusion would be inescapable: flying these missions in the Keystone Puffer was tantamount to a death wish. By then, however, the spring thaw had begun breaking winter's icy grip on Canada in 1928, rendering time of the essence. Forest pests of all sorts would soon be waking up and beginning another season of feasting on various species of valuable trees. As a result, the rush was on to find a safer plane to deliver the poison. Overcoming this hurdle would allow the Canadian campaign to become more proficient at aerial dusting over forests to move into its next phase, a subject to which we now turn.

CHAPTER THREE

"Fighting insect plagues is something new"

AERIAL DUSTING FOR INDUSTRIAL FORESTRY
IN ONTARIO AND QUEBEC, 1928–1929

Malcolm Swaine, Canada's chief forest entomologist, was determined to build on the momentum his experimental aerial spraying campaign had generated in Nova Scotia in 1927, and to do so his focus shifted to several new theatres of operation. The first was in northern Ontario, where he targeted the spruce budworm in 1928 and 1929. The second was in the Baie Comeau region of Quebec in 1929, where he aimed to turn back an infestation of hemlock looper. Like the effort against the budworm in Nova Scotia, these endeavours were motivated by a desire to assist the country's forest industry.

The dusting experiments in central Canada also mirrored the trials in Nova Scotia in terms of the dynamics that drove them and the issues that surrounded them. Particularly noteworthy is the extraordinary measure of cooperation that, like the effort in Nova Scotia, likewise defined these projects. These efforts in Ontario and Quebec saw officials from different companies, several provincial governments, and a number of branches of the national bureaucracy working together to realize their common aims; the dusting projects reflected the synergistic spirit that defined forestry in Canada during this period. Moreover, the firms that eagerly embraced these dusting projects did so because they had long demonstrated a fervent commitment to and faith in adopting the latest science in their industrial activities.

There were other similarities as well. Nature continued to play a dominant role in influencing where, when, and how the entomologists conducted these projects. The weather, for example, imposed its own timetable on the efforts. In trying to kill both the spruce budworm and the hemlock looper by dropping dust on them from airplanes, scientists also came to appreciate that these two forest insects were as different as night and day. Although the two species were roughly the same size and went through the same basic life cycle, their particular biological characteristics imbued each with vastly different intrinsic safeguards against the toxic powders. Nature also shaped the ebb and flow of the particular epidemics that these projects aimed to address, specifically through the activities of parasites. These dusting projects in Ontario and Quebec over the course of 1928–9 also displayed once again how Canadian pilots and scientists were true pioneers in conducting this type of work. The former flew their missions in the face of remarkable danger, and the latter repeatedly improvised solutions to the string of practical problems that arose to confront them. All the while, the foresters and entomologists continued to insist that the best and most lasting solution to the forest pest problems they faced was the most environmentally friendly one: namely, adjusting harvesting techniques to render the woods less vulnerable to insect depredations.

Concern about the spruce budworm was nothing new to the forestry community in Ontario and Quebec. Since at least the end of the First World War, forestry officials in central Canada had been growing increasingly worried about the threat the budworm represented to the region's commercial woodlands, with many opting to suggest a natural solution to the problem. One of the leading spokespersons in this regard was Clifton Durant Howe, the dean of the Faculty of Forestry at the University of Toronto (1919–41). Under the auspices of Canada's Commission of Conservation (1909–21), Howe had been involved in numerous trailblazing investigations into the state of the country's woodlands during the 1910s. These experiences had laid plain to him both the spruce budworm's life cycle, and the best means of thwarting the damaging effect it was having on the region's trees. After examining the devastation the budworm had wrought in New Brunswick and Quebec, Howe realized that balsam fir was its favourite food, and that this species rarely survived successive years of attacks. Howe thus concluded in the early 1920s that the best approach to mitigating the destruction the budworm caused was to concentrate the industry's harvest on mature and over-mature balsam fir, and to keep the age of this species as young as possible in the residual stands. This was precisely the prescription dominion entomologists such as Malcolm Swaine had begun recommending around this same time.[1]

Canada's largest pulp and paper company was even more concerned than Howe about the spruce budworm and was particularly receptive to taking steps to combat it. The Spanish River Pulp and Paper Company operated mills in Sault Ste. Marie, Espanola, and Sturgeon Falls in northeastern Ontario (Map 1). The firm's president, George H. Mead, was an ardent proponent of Taylorism, a theory of management that suggests a business can maximize its efficiency only by applying scientific principles to all aspects of its operations. Spanish River had thus been at the forefront of Canadian companies in terms of adopting "modern" approaches to human relations, for example, and in launching a research and development program. It had also been eager to apply this business ethos to the management of its woodlands, and to oversee the process it had hired a dynamic American forester, Benjamin F. Avery, during the First World War. Within short order, Avery had implemented a comprehensive forest management program based upon the concept of sustained yield, and the strategy's critical first step was to investigate the makeup of Spanish River's timber holdings. This entailed conducting aerial surveys and ground cruises of the tracts the firm leased from the Ontario government. While carrying them out, it became glaringly apparent that the spruce budworm was alive and well in Spanish River's woodlands.[2]

Initially, Avery was convinced, just like Howe and Swaine, that the best means of dealing with this insect pest was to harvest older balsam fir. Over the course of 1920–1 he drew up a working plan (i.e., the strategy for managing a tract on a sustained yield basis) for the Goulais River watershed just north of Spanish River's mill in Sault Ste. Marie. The Commission of Conservation had conducted a regeneration study in this area the year before and had thus created the foundation of data upon which Avery's plan could rest. It underscored the vulnerability of the balsam fir to the spruce budworm's attack and made a compelling case in favour of decreasing its presence in the woodlands controlled by Spanish River.[3] A few years later, his annual report to the company reiterated this recommendation in light of the spruce budworm's continued attacks.[4] By the mid-1920s, not only was he continually exhorting his woodsmen to follow this directive but so too were Ontario's other leading pulp and paper firms.[5]

A few years later, Spanish River had even greater reason to be troubled by the presence of the spruce budworm in its woodlands. In July 1927, the company had hired Frank T. Jenkins, one of Canada's pioneer aerial sketchers and timber surveyors, to investigate its timber limits north of Sudbury. The report he submitted to the firm delivered highly disturbing news. Jenkins described how "[a] serious budworm infestation [wa]s threatening to wipe out not only the balsam but also the spruce on a very large part of this area. The trees were just beginning to turn red when this survey was made but indications point to a much more general infestation than was at first thought ... Many of the best stands of pulpwood have been most seriously infested and logging operations should be planned so as to remove all injured trees within the next few years." Jenkins indicated that "the most seriously infested" area was centred near the town of Westree (Map 1).[6]

Around the same time, Malcolm Swaine and the dominion forest entomologists began providing corroborative accounts that turned gloomier by the month. In mid-1927 they reported that the spruce budworm was "heavily infesting balsam fir and spruce" in the same location, within which nearly all the former was dead and the latter severely damaged. The budworm's appetite was so voracious that it had even started feasting on the local white pine and larch.[7]

Avery was thus keen to take definitive action. Soon after receiving word of Jenkins's findings, Avery dispatched A. Harold Burk, Spanish River's Assistant Forester, to investigate the situation further. By late winter 1928, Avery had confirmation that the situation was grave. Burk's visit to the affected area revealed that, in at least one entire township (thirty-six square miles), "[t]he Balsam is dead, 100%, even to the smallest tree. The larger spruce has all been hit." Burk considered this an ideal opportunity to learn about the budworm, which, he noted, had attacked in a very irregular pattern; pockets of balsam fir and white spruce had survived within areas that were generally infested. He thus recommended carrying out an aerial survey of the affected tracts in order to determine the extent of the outbreak, information that would facilitate possible future control work.[8]

Moving forward, Avery was keen to find partners with whom to cooperate, and he turned first to the provincial government. Beginning in mid-February, he began forwarding reports on the budworm infestation to Edmund J. Zavitz, Ontario's Deputy Minister of Forestry. Avery reminded Zavitz that "the budworm is doing great damage to the spruce and balsam north of Sudbury," and that measures were urgently needed to counter this threat. As an initial step, Avery passed along Burk's recommendation to undertake an aerial survey of the infestation as well as Burk's offer to do this work himself if the Ontario government supplied the plane. In this regard, Avery believed that "the control of this insect damage, is a Provincial function, but since this Company is suffering loss through the damage, we shall be glad to co-operate with you in any undertaking to control or stop the spread of this insect pest." Avery thus suggested that Spanish River and the Forestry Branch of the Ontario Department of Lands and Forests (DLF) carry out the preliminary reconnaissance together. Looking ahead to actually combatting the problem, Avery pointed out that he had learned that "dusting experiments" had been carried out down east, "and that the reports on them had indicated their effectiveness to some degree."[9]

The Ontario government received Avery's offer with enthusiasm, particularly the DLF's officials in the area of the budworm infestation. In mid-February 1928, Kelvin A. Stewart, the DLF's District Forester in Sudbury, advised Zavitz that he eagerly supported Avery's strategy of having the boundaries of the infested area defined by an aerial survey. Stewart emphasized that Spanish River had committed to the project even though the budworm outbreak extended far beyond the firm's timber limits, and he thus urged the Ontario government to help finance this exploratory operation.[10]

Fortunately for Avery and Spanish River, a conjunction of interests pushed along the work on this project in an efficient manner. As we know, Malcolm Swaine had been eagerly searching for an opportunity to conduct a large-scale aerial dusting experiment against the spruce budworm that season, and by now he knew that there was no hope of conducting another project in Nova Scotia because that province's budworm infestation was declining of its own accord. By early 1928, Swaine and Avery had agreed that both their interests would thus be served by carrying out a trial dusting operation north of Sudbury. Swaine wrote Zavitz in early March to inform him of this plan and to ask him to help coordinate the effort.[11] Zavitz agreed and set out to uphold his government's end of the bargain. He convinced William Finlayson, Ontario's Minister of Lands and Forests, to ask the dominion government for permission to borrow the same aircraft that the Air Board had employed in the dusting experiments the previous season in Nova Scotia, the Keystone Puffer.[12]

It was at this point, unsurprisingly, that the effort began to founder. As we have seen, ever since the dominion entomologists had first floated the idea of experimenting with aerial dusting against forest insects, officials from the Air Board had expressed their deepest reservations about such projects because of the acute dangers involved. Even before the dusting trial in Nova Scotia had been carried out, the evidence to substantiate the Air Board's unease had been piling up.

At the top of the mound was a report from Flying Officers Bath and Shields. It will be recalled that Bath and Shields had been tapped to fly the inaugural dusting missions back in mid-1927, and that at that time their superiors had ordered them to Tallulah, Louisiana, to be trained by Bert Coad in the techniques of aerial dusting. After their stay in the Deep South, during which inclement weather had drastically curtailed their flying time, Bath and Shields had concluded that their lack of training in the cockpit of the plane was immaterial because the plane – not its pilots – would be the biggest impediment to the Canadian dusting experiments. The pilots were convinced that the Keystone Puffer was unsuited to dusting over commercial woodlands because of its lack of maneuverability. As Bath put it, the machine "was absolutely unstable and required very careful attention at the low height necessary for dusting. It also has a tendency, when zoomed sharply, to avoid trees at the end of a field, to continue travelling in the horizontal plane for a short distance." Bath also found it more than a little unsettling that the plane had been purposefully manufactured a certain way because of the high probability of crash landings in this type of work. As Bath described it, the aircraft was "very strongly constructed between the engine and the pilot, to eliminate as far as possible the danger of injury to the pilot in the event of a crash."[13]

The situation was exacerbated because of a central flaw in how the Canadians had asked for their particular Puffers to be configured. When the dominion officials had first contemplated buying the aircraft for dusting work back in early 1927, they had sought to spend as little as possible, but achieving this aim would dramatically increase the danger involved in flying these dusting missions. It was decided by officials from both the RCAF and the Division of Entomology that they would need a pair of planes, one for the forest entomologists and another for the agricultural entomologists (the latter were keen to oversee trials that used poison dusts to kill rusts that were attacking grain crops in the Prairies). They also shared the view that the best plane for the job was the Keystone Puffer, which had been designed specifically for dusting work and whose performance had been stellar over the open fields of the American Cotton Belt. But the plane's hefty price tag meant that, even with the support of James Ralston, the Minister of National Defence, the Air Board could only acquire the planes if they could be converted to other uses when they were not needed by the entomologists for dusting.[14] To meet this condition, the planes were outfitted with pontoons and their fuselages were modified. The planes were also customized to allow the hopper, the device into which the insecticide dust was loaded and from which it was dispensed, to be removed and replaced by other equipment, such as cameras for aerial photography, when the plane was not being used for dusting. The Puffer's chief engineer assured the Canadian officials that the plane could be customized in this manner, but that "it would be best to leave the rudder bar and elevator jack shaft in place."[15] The Canadians ignored this advice, however, and insisted that these devices be moved.[16]

Flying Officer Bath had been forced to deal with the perils of this decision when he had flown the missions in Nova Scotia in 1927. The problem was that this arrangement had resulted in the hopper being positioned "directly under the centre of lift," and to compensate for this realignment of the plane's cargo load the controls for adjusting the tail wing should have been located adjacent to the pilot's seat. But the Canadians had not known about this consideration when they had ordered their custom-built Puffers. As a result, when Bath asked the plane's manufacturer why the firm had not made the tail adjustable from the pilot's seat, the officials at Keystone answered rather matter-of-factly that "it had not been asked for in the specifications."[17]

Bath's experience in conducting the dusting operations in Nova Scotia had convinced him that his concerns about the Puffer were profoundly warranted. As discussions about the projects planned for 1928 were heating up, Bath sent James Lindsay Gordon, the Air Board's director, a detailed report about his experiences and reservations regarding flying the Puffer during the forest dusting operations in Nova Scotia. Bath's conclusion was unequivocal – the Puffer was utterly unsuited for this type of work:

> As one is necessarily flying as low as one can possibly fly over the tops of the trees, I consider that this work is too dangerous to use a single-engine machine. If anything should happen to the engine one would have to land on the trees and the pilot would have a very slight chance of surviving. This is especially true when using the Puffer, as with floats and 600 lbs. of dust she starts to sink at about 75 miles an hour. I should think that even an H.S.2.L [i.e., a model of plane that had been commonly used in bush flying since the First World War] would be better as she has a much greater wing surface and can be pancaked on to the trees at a much slower speed.[18]

In addition, Bath pointed out the practical problems inherent in using the Puffer for this type of work, specifically how its relatively small payload made it highly inefficient as a forest-dusting plane. He recounted how it took only a minute and a half to dump 600 pounds of dust over the trees, but took between thirty and sixty minutes to reload the plane with that same amount of toxin for a second run.[19]

For good reason, Bath's report carried enormous weight and resonated strongly with his superiors at the Air Board. Moreover, everyone involved in the inaugural dusting effort in Nova Scotia had concurred on one thing – that the effort had, quite literally, only gotten off the ground because of Bath's superb skills as a pilot. As a result, when Air Board Director Gordon had asked Wing Commander Ernest W. Stedman, the Chief of Aeronautical Research and Development for the RCAF, to comment on Bath's report, Stedman accepted it unquestioningly. "There is no doubt about the danger of dusting over tree-tops, or of the economy of using a larger machine if large areas have to be dusted," Stedman wrote. He thus recommended using a three-engine aircraft in the future to decrease the hazardous nature of this work.[20]

But all this talk of and concern about the extreme danger involved in using the Puffer for forest dusting work seemed to matter little.[21] The Air Board's officials recognized this grim reality, but still decided to go forward with these projects using the much-maligned plane. At this point, they were broadcasting near and far the wide range of services the Air Board was providing to Canada's natural resource sector and abhorred the very hint of any negative publicity surrounding their work. As a result, when Stedman delivered a speech to the Engineering Institute of Canada in Montreal in February 1928, he said nothing about the dangers involved in bombing forests with chemicals and instead stressed – using hyperbole befitting a salesman – that the spruce budworm "was being conquered by a system of dusting forests from the air."[22]

It was much the same story with senior officials from Canada's Department of National Defence (DND), who admitted the Keystone Puffer's failings but still endorsed its use that year. In early April 1928, the Air Board's Director James Gordon had informed George J. Desbarats, the Deputy Minister of the DND, that the Puffer was not appropriate for this type of flying and strongly suggested using a "multi-engine type" instead. Desbarats arranged a meeting with Joseph Grisdale, the Deputy Minister of Agriculture, to discuss the matter, specifically because the air force officials believed that "the risk to the personnel and existing equipment engaged in an operation of this kind, is considerable."[23] Nevertheless, by the end of April, Desbarats had arranged for a modus vivendi with the entomologists that did little to safeguard the pilot's safety. Indeed, his communications with them reiterated that dusting forests far from landing facilities with the Puffer entailed "considerable risk to the personnel and equipment."[24] Gordon had been even more candid in describing how "flying of this kind is exceedingly dangerous and in the event of engine failure, especially prior to the load having been distributed, a serious crash would be inevitable."[25] But all the same, Desbarats agreed to use the plane that season to conduct two dusting operations in Ontario (one against the spruce budworm north of Sudbury and another in the Muskoka District, about which more will be said in the next chapter). All he asked in doing so was that the work proceed "subject to the *possibility* [emphasis added] of choosing test plots on islands, promontories, or close to edge of lakes where, in the event of engine failure, it would be possible for the pilot to alight safely on the water." Desbarats added that the DND was hoping in the near future to purchase a new plane to conduct these missions, one that had "two or three motors in order to increase the safety factor and enable the aeroplane to dust forests more remote from water areas."[26]

In the meantime, the participants in the project forged ahead in a manner that was defined by a spirit of cooperation and pioneering. In mid-May 1928, Harold Burk, Spanish River's Assistant Forester, carried out the aerial reconnaissance of the "areas known to be hit by the budworm" in the Sudbury District in an effort to locate specific sites that would be best for the trial.[27]

Shortly thereafter, Spanish River sent ground crews into the area (assisted by personnel from the dominion's Division of Entomology and Ontario's DLF) to identify suitable standard "sample plots" and to record conditions prior to the

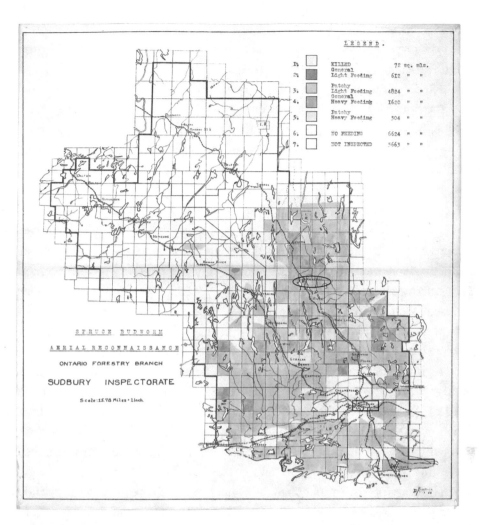

3.1 Map illustrating the spruce budworm infestation on Spanish River's woodlands, 1928. For reference, Sudbury is in the lower right of the map, with a rectangle around it. Westree is circled and located on the railway line that runs north-north-west of Sudbury. The original map is in colour. On this black and white version, in general the darker shading represents areas in which the spruce budworm infestation was most severe. Source: SMPA, A-1, Forestry 1928, Report of reconnaissance flight made for the purpose of locating sample plots for experimental dusting of budworm infested timber, enclosed in 15 May 1928, A.H. Burk to J.M. Swaine.

application of the chemical dusts. These reports attest to the devastation the budworm had wrought. They repeatedly describe stands in which both white spruce and balsam have been heavily hit and the "[b]alsam is almost stripped of leaves to ½ the height of tree. Spruce has last year's needles stripped for the same distance."[28] Having received these reports, Malcolm Swaine commended Burk for the "very excellent" information Spanish River was generating and for the extraordinary esprit de corps that was evident among the numerous participants in the project. Swaine proclaimed that he was "exceedingly gratified that we are going to have this excellent cooperation among our four organizations [i.e. Spanish River, the Ontario DLF, the dominion's Division of Entomology, and the DND's Air Board]. This will give us the best possible opportunity for obtaining valuable results."[29] In the end, Spanish River concluded that the village of "Westree will offer best possibilities for establishment of an aeroplane base for the experimental work" (Map 1).[30]

This same symbiosis pervaded the final preparations for the project, which were made in late May and early June. Swaine and Flying Officer Bath – the latter was now being referred to as "Officer in Charge of Dusting Operations" – worked with personnel from Ontario's DLF to select the nine plots to be dusted along the shore-line of two large waterbodies, five on Duchabani Lake and four on Donnegona Lake. Each one measured roughly 500 feet wide by 1,000 feet long and was chosen to represent the different forest types affected by the budworm outbreak. In making their preparations, Swaine and Bath also applied the knowledge they had gained the previous season in Nova Scotia by tying windsock-like pieces of cloth to trees that were highly visible from the air. Meanwhile, the DLF's men established the bush camp at Westree that would serve as the base for the dusting operation.[31] On 6 June, Swaine informed Zavitz that everything would be "ready for dusting by the time the buds and caterpillars are in the right condition. Judging by the rate at which the buds are opening now, the dusting will probably be done some time next week, the date depending on the weather." Again, Swaine was grateful for the remarkable degree to which the Ontario government had facilitated setting up the project. He confided to Zavitz that he was "very much pleased indeed with the splendid co-operation you are giving us," and that he felt that "everything possible has been done to ensure a thorough test of this method."[32]

Mother Nature, however, would wreak havoc on the project soon after it began. In fact, Swaine considered the effort such a write-off that he never filed an official report on it. Likewise, the Air Board was so reluctant to publicize its involvement in the project that its officials barely mentioned the Westree operation in the annual report they filed for 1928.[33] Swaine and his partners in the effort had set out several project objectives, including determining the minimum dose of calcium arsenate needed per acre to kill the budworm, the most effective brand on the market, and how best to conduct such operations. It initially appeared that they would begin realizing these aims when the first dusting flight took off on 17 June. Immediately thereafter, however, the problems began, with "unfavourable weather" delaying the next sortie. While the dusting was on hold, Flying Officer Bath carried out test flights to learn how best to regulate the rate at which the hopper discharged the dust, but the hopper soon broke and had to be repaired, causing further delays.

The challenges continued for Swaine and his crew. When the weather improved enough over the course of 22–23 June and the operation recommenced, heavy rain started falling soon after Bath had applied the chemicals. As one observer put it, this "probably washed all the dust off." Then after the heavy rain came several days of high winds. When they died down enough to begin flying again, Bath dumped another hopper of the calcium arsenate dust. Upon returning to the base at Westree to reload, however, Bath damaged the plane's pontoons and knocked it out of action, necessitating that the plane be flown to Ottawa for repairs. After this latest setback, Kelvin Stewart, the DLF's local senior forester, reported on the immense difficulty that was being encountered in getting the "almost absolute calm" needed to carry out the dusting. This had become a particularly pressing issue because the caterpillars would be, as he put it, "in condition to dust" for only the next seven to ten days. While Stewart wondered what Swaine would do if time ran out before the dusting was completed, he noted that Swaine had ordered a number of "hand spray pumps" as a back-up plan.[34]

At this point, all concerned tried to make the best of a bad situation. When the weather still refused to cooperate, Aubrey MacAndrews, Swaine's colleague who had been in charge of much of the fieldwork at Westree, began using the hand sprayers Swaine had ordered in an effort to learn more about the manner in which the budworm reacted to different concentrations of dust. Ultimately, this seemed to have been the most valuable part of the entire enterprise, as the Puffer was able to fly for a total of only ten hours over the project's three weeks' duration. Harold Burk, Spanish River's Assistant Forester, aptly summarized the situation in a letter to his superior at Spanish River, Chief Forester Ben Avery: "The main object of the experiment was lost, namely the economic possibility of airplane dusting. However, the biological results obtained by hand dusting will be of some use."[35]

These events were certainly disappointing for all concerned, but the budworm's activities were about to cause even more grief. In mid-July 1928, Burk provided Avery with an update on the infestation north of Sudbury. He described how he had flown over the infestation with officials from the Ontario Forestry Branch and mapped the approximate boundaries of both the old and newly infested areas. "As you will note," he told Avery, "the migration was of very serious proportions, so much so, in fact that any control work is now entirely impossible." He thus despaired that "we will have to let it run it's [sic] course, then concentrate control operations on any small areas where there is a hang-over such as took place after the last outbreak."[36] The dominion entomologists likewise confirmed that the situation was dire through their monthly reports on Canada's most pressing pest issues.[37]

Next, it was the Air Board's turn to deliver even more dismal news. Flying Officer Bath was sticking by his earlier grim assessment of the Puffer's suitability for this type of work, and R.S. Grandy, Bath's Squadron Leader, concurred. Grandy reported to the DND's secretary that aerial dusting over forests was "considered to be extremely dangerous," and that the Puffer was unsuited for these types of flying missions. "The Keystone Puffer with a full load of dust lands at approximately 70 to 75 m.p.h. and it can be readily imagined what would happen if a forced landing occurred when flying over the bush a few feet high as they have to. From observations made it would seem that this forest dusting is not very practical," he concluded. "As the areas are so great and the conditions have to be so ideal it is doubtful whether enough territory can be covered to make it worthwhile." Reading between the lines, the message was clear. The Air Board's personnel would not fly another dusting sortie using the Puffer over Canada's forests.[38]

Although Swaine was acutely aware of this situation and the other challenges his aerial dusting campaign faced, he remained steadfast; to him, the hopper was always half full rather than half empty. To all concerned, he expressed his gratitude for the work that they had carried out against the budworm at Westree and his optimism about the projects he was already planning for the following season. Just as the one near Sudbury was wrapping up, for example, he wrote James Gordon, the Air Board's director, to acknowledge Gordon's help. In the same breath Swaine both lamented the inclement weather that had hampered the effort's progress that season and downplayed the near miss that Flying Officer Bath had had with the Puffer. "I am very sorry," Swaine wrote Gordon, "about the accident to the plane and we are all glad that it was not more serious."[39]

Swaine was equally tactful in dealing with officials from the Ontario government. "The unfortunate weather conditions and the crash to the plane," he told Kelvin Stewart, the Ontario government's District Forester for Sudbury, "have unfortunately prevented us from obtaining any definite results from the airplane dusting operations." At the same time, he highlighted the silver lining to the clouds that had practically grounded his dusting effort. Swaine stressed the valuable data that MacAndrews had derived from the hand dusting operations. In particular, he stressed how MacAndrews had demonstrated that, with sufficiently heavy doses, "the caterpillars can be killed even within a few days of pupation," and that from this effort they had obtained "very useful data on the biology of the insects in that region." Furthermore, Swaine explained that he and his colleagues intended to continue the project on a larger scale the next year "so that the dust may be applied even in moderate winds." He continued: "We can have a few small plots ready for testing the doses more definitely, to be used if weather conditions permit; we shall not again take the chance of losing so much by depending on the perfect weather conditions that are essential for treating small plots. It will involve using large quantities of dust, but that seems unavoidable." Swaine closed his letter to Stewart by reiterating his "deep appreciation for the perfect co-operation we have had from your organization, in this Westree work," and signed off looking forward to the future: "Hoping to see you and Burke [sic] [i.e., Spanish River's representative] in Ottawa this winter, according to the plan we discussed while we were at Westree."[40]

But Swaine was putting the cart before the horse, which was especially ill-advised in this instance since he knew that the horse itself was lame. The DND's experience with the Puffer during the dusting project at Westree in mid-1928 – and during another effort that occurred later that summer (see chapter 3) – reinforced a point that everyone already knew. Flying this aircraft to dust forests was likely to be fatal to the pilot, and the DND was no longer willing to roll the dice with its airmen. George J. Desbarats, the DND's Deputy Minister, relayed his position to Joseph Grisdale, his counterpart in the Department of Agriculture, in late 1928. In asking for Grisdale's wish list in terms of prospective dusting projects for the following year, Desbarats was adamant that he would gladly continue using the Puffer to carry out the trials designed to combat wheat rust in the Prairies. But if Swaine was hoping to conduct more "forestry dusting" efforts with the same aircraft, Desbarats declared defiantly, "a new type of airplane should be considered which will not involve undue risk to the personnel engaged on the operation."[41]

Nevertheless, Swaine would not be dissuaded, and with good reason. Spanish River was still eager to continue dusting the budworm infestation that was feasting on its timber holdings even though it sensed that this battle had already been lost. It welcomed any measures that would help conserve its fibre supply because it was already grappling with a wood shortage.[42] In addition, Ontario's Forestry Branch was equally committed to pushing the campaign against the budworm forward while helping Swaine learn more about how best to combat the insect using aerial dusting techniques. By the end of 1928, the deal was sealed. Westree would serve as the base for another campaign against the budworm the following season.[43]

Moreover, Swaine now had a new request from a different player in central Canada's forest industry for help in fending off an insect attack. It involved the eastern hemlock looper (*Lambdina fiscellaria*) in Quebec. Like the budworm, this pest's name is a misnomer, for it gladly feasts on balsam fir and black and white spruce if hemlock is not available nearby. The adult is a small moth that has thin, grey to cream-coloured wings that are scalloped with darker bands. It usually emerges in August and takes flight for mating and egg-laying the following month. It lays its tiny, light green eggs on the twigs and foliage of the host tree (or on rotten stumps), which turn to brown before hatching as slender green to nearly black caterpillars the following June. The caterpillars walk with a looping motion, thereby giving them their common name, but they are also called "span-worms" or "measuring-worms." They feed on conifer needles for roughly six weeks and then become pupae in August. During this resting stage they are found under loose bark on the lower trunks of conifer trees or under litter on the forest floor. After about two weeks the pupal skin splits and the moth emerges. Looper infestations typically break out with lightning speed, continue for a few years with acute intensity, and then dissipate just as quickly as they had started. They typically kill only those trees that are defoliated each year of the attack.[44]

Although the dominion forest entomologists had already been tracking a relatively massive looper infestation in Quebec's Saguenay River valley, a handful of much smaller ones had also started to draw their attention. These invasions were scattered in the Manicouagan River valley, which fronted onto the St. Lawrence River's north shore in the region of Baie Comeau (Map 1). The Ontario Paper Company (OPC) and the Anglo-Canadian Pulp and Paper Company co-managed large swaths of timber in this watershed. The Tribune Company had created OPC in 1912 through which it constructed a newsprint mill in Thorold, Ontario, to supply paper for its eponymous daily in Chicago. From the outset, OPC incorporated the latest science and technology into its enterprises, including when it built the province's first pulp and paper mill driven entirely by electricity instead of direct water power, and later when it began to turn formerly waste wood by-products into valuable commodities such as ethanol and vanilla.[45] As a result, OPC was a prime candidate for participating in an aerial dusting operation.

As luck would have it, OPC and Anglo-Canadian's foresters had discovered in the summer of 1928 isolated looper outbreaks scattered on the woodlands they controlled and forwarded their concerns to the Quebec Forest Industries Association. Its manager had taken the matter up with Swaine and his colleagues, and soon they were cooperating with the Quebec Forestry Service to investigate the situation and discuss the possible means by which they could address it. In the process, they located a particularly heavy infestation at the mouth of the Manicouagan River that they feared would spread into the stands of practically pure balsam fir that stretched northward up the watershed. In short order, it was agreed that an experimental dusting project would be launched against the "Manicouagan outbreak" of the hemlock looper the next season.[46]

As Swaine revealed his dusting plans for 1929, their audaciousness betrayed his indifference to the challenges they faced. He knew that the Air Board would not carry out these missions using the Puffer and that the air force officials had still not acquired a suitable alternative plane. Nevertheless, whenever he discussed the operation he hoped to conduct at Westree that year, its scope was exponentially greater than any effort he had ever carried out. Writing to Edmund Zavitz, Ontario's Deputy Minister of Forestry, about the Puffer in February 1929, Swaine admitted that the plane had been too small for their purposes and that it had been "entirely too dangerous to fly that unstaple [*sic*] machine close over the tree tops." At the same time, he outlined how he intended to "continue the Westree work but on a much larger scale." Specifically, he wanted to apply approximately twenty tons of calcium arsenate over roughly two square miles of timber, with the project's overriding goal to prove that this type of work was economically viable. "If we can kill the caterpillars successfully with 25 pounds per acre," Swaine explained, "this method should prove of inestimable value, and it should be possible to arrange extensive dusting operations to check the spread of outbreaks such as that in the Westree district at an expense that could be borne." In the same breath, Swaine asked the Ontario government to pay for the dust that would be required and to provide the same ground support that it had made available the previous year, requests that Zavitz eagerly agreed to meet.[47]

While Swaine was planning the projects, officials from the Air Board were tackling the issue of finding a new aircraft suitable for dusting forests, and were again being forced to improvise for the sake of the projects being able to move forward. During the spring of 1929, they received unsolicited offers from a German aircraft manufacturer – the Junkers Corporation – for dusting aircraft, and they understood the allure of the sales pitch. At this time, the patent holders for aerial dusting hoppers refused to sell their equipment separately from the planes that used it. This meant that if the Canadians did not purchase a dusting aircraft like the one the Junkers Corporation was offering, they would have to design and build their own hopper apparatus. The Air Board chose to pursue the latter option that May when it decided to purchase a trimotor Ford 4-AT – the first three-engine aircraft in eastern Canada – and asked the manufacturer to install a hopper in it.[48]

In the meantime, Swaine's aerial dusting campaign was garnering increasing domestic and international attention. At home, the Canadian media's interest and pride in the nation's innovative work in this field was swelling, as was its use of battlefield rhetoric to describe the efforts. In late January 1929, Toronto's *Mail and Empire* lauded Canada's "Intensive War to Save Timber" from the ravages of insects (and fire). The article chronicled how the experiments over the past few years had demonstrated that "the insect pests" could be "eradicated to a large extent by spraying the infected forests from specially-equipped air machines." Likewise, the *Border Cities Star* in Windsor, Ontario, provided detailed coverage of the delivery of the trimotor Ford aircraft to the Canadian government and the dusting work the plane would soon be undertaking against the spruce budworm.[49] More remarkable yet was the notice that aviation buffs around the world were taking of the aerial campaign against forest insects that Swaine and the Canadian entomologists were waging. An article published in mid-1929 in the international *Flight Magazine* rhymed off Swaine's accomplishments, as well as the success enjoyed by those who were using planes for other peacetime purposes in Canada. In fact, the piece even caught the eye of a Japanese student in an agricultural and forestry school in Tokyo. He was so intrigued by what he had read that he was compelled to write to the Canadian officials to explain that "it was a great surprise ... to know the wonderful progress existing in your country" in terms of aviation's industrial applications.[50]

By the late spring of 1929, Swaine was keen to continue making this "wonderful progress" both on the ground and in the air. In terms of the project to combat the spruce budworm outbreak north of Sudbury, he ensured that the preparations were well underway by late May. At that time, personnel from the Ontario Forestry Branch (OFB) re-erected the bush camp near Westree on the shores of Duchabani Lake. It had to be slightly expanded in order to accommodate the crew from the Motion Picture Bureau – the predecessor to the National Film Board (NFB) – that the dominion government sent to the site to capture the activities for posterity; the product was a short, silent documentary, which survives in the NFB's archives.[51] In addition, the OFB's staff helped cut the dusting plot boundaries after dominion entomologist M.B. Dunn had conducted a preliminary investigation to determine the best sites to treat. Another task they completed was constructing an improvised service station for the aircraft. Made out of spruce poles, the twenty-five-foot tower was used to refuel the plane. By mid-June, the trimotor Ford aircraft and its pilot, Flight Lieutenant N.C. Ogilvie-Forbes, had also arrived on the scene, as had Malcolm Swaine, who would direct the project.[52]

The new flying machine hardly turned out to be the panacea for which the entomologists and Air Board officials had been hoping; their experiences at Westree that season gave new meaning to the axiom that necessity is the mother of invention. In fact, even as Ogilvie-Forbes had been flying the plane from Ottawa to Westree, he had been forced to make an emergency landing because of a "serious oil leak" and continue the journey on only two engines. On 15 June, the day the plane arrived in Westree, the airplane's crew lost its starting crank in the lake and then realized that its electric generator was not working. They fixed the latter problem and

3.2 Tower built for refueling the plane at Westree. Source: NFB, Shot ID1695, An aeroplane dusting experiment for the control of spruce bud worm carried out near Westree on [*sic*] June 1929.

improvised a crank constructed from gas piping to resolve the former. When the dusting commenced two days later, Ogilvie-Forbes described how all the engines had suffered from "serious overheating," prompting the airplane's engineers to ask the Ford Motor Company for help in resolving the issue. Thereafter, the middle engine continued experiencing "oil pressure trouble," and Ogilvie-Forbes asked for permission to fly the crippled plane back to Ottawa for a complete overhaul. His superiors rejected his request, however, because time was short. It was already mid-June and the trimotor Ford was expected to begin dusting the hemlock looper outbreak in Quebec by 1 July. As a result, the ground crew did all it could to keep the plane airborne until the project ended on 27 June.[53] In addition, the hopper proved problematic on practically a daily basis. Initially it delivered a dusting that was far too thin to be effective. It took a few days to overcome that hurdle only for the hopper to then break down completely. By the middle of the dusting operation, however, Swaine could triumphantly report that "the hopper trouble had finally been overcome and an excellent dust cloud obtained."[54]

Although the project also endured unfavourable weather – heavy rains frequently fell shortly after the plane had dusted the plots – in general it proved far more successful than the one conducted in the same location the previous season. Over the course of the last few weeks of June 1929, the trimotor Ford was able to drop more than 35,000 pounds of calcium arsenate over roughly 1,500 acres of infested forest. The ambit of the project thus enabled it to realize its primary goal, namely to conduct an operation on a commercial scale in order to generate solid data regarding

3.3 Close-up shot of the Ford Trimotor dusting the spruce budworm infestation near Westree. This remarkable photograph was taken from another aircraft that flew just above and behind the Ford. Source: NFB, Shot ID1695, An aeroplane dusting experiment for the control of spruce bud worm carried out at Westree on [*sic*] June 1929.

its costs.[55] In addition, the entomologists compared their untreated test areas to those that were doused with the poison, and their results showed that the dust had caused immediate mortality among the budworm larvae and also continued to kill them for about two weeks. They recognized that parasites were also knocking back the infestation, but Swaine asserted that the mounds of dead budworm caterpillars were overwhelmingly attributable to the dusting operation. "It is probable," he thus contended, "that the same treatment applied to a budworm outbreak in an early stage, with the new engine and improved hopper employed in the subsequent dusting operation in Manicouagan [discussed below], would have given a satisfactory degree of control with a least 30 and 40 pounds of dust per acre."[56]

But Swaine's desire to squeeze better results from the data and portray the campaign against the budworm in a positive light could not overcome one glaring problem: conducting these aerial campaigns against this insect in central Canada's pulpwood forests simply did not make economic sense. A few years after the Westree operation concluded the entomologists had lowered their estimate of its

3.4 Distant shot of aerial dusting near Westree. Source: NFB, Shot ID1695, An aeroplane dusting experiment for the control of spruce bud worm carried out at Westree on [*sic*] June 1929.

effectiveness to about 70 per cent. That was not, as the major daily in Fredericton noted, "nearly high enough in the opinion of the scientists who undertook the work" to justify the treatment.[57] The fundamental problem was the relatively low value of the pulpwood forest that was being protected.[58]

Fortunately for those worried about the budworm outbreak in the Westree area north of Sudbury, their fortunes were once again enhanced by natural forces that must have seemed like divine intervention. In December 1930, Harold Burk, a senior forester with Spanish River (now Abitibi Power & Paper), sent Ben Avery, Abitibi's Chief Forester, a map that indicated the extent of the budworm outbreak in the Sudbury District as of October 1929. "I have reason to believe," Burk surmised, "that the increase in infested area was small this year, and although larger areas of balsam will be found to be entirely killed the feeding in general was much lighter than it has been for some years."[59] Likewise, after conducting an aerial survey of the affected zone a few years later, Abitibi's foresters noted "that the areas of infestation have not spread during the year and in fact large areas which showed light feeding in 1931 appeared to be free from infection in 1932. The extremely rapid spread of the budworm seems now to have been checked by natural causes of parasitism and possible adverse weather conditions and the area of heavy feeding is greatly reduced."[60]

But the story would be very different with regard to the outbreak in the Baie Comeau region of Quebec, where the pressing forest pest had been the hemlock looper and balsam fir its tree of choice. Swaine pursued his plan to conduct a major aerial dusting operation there with unprecedented vigour. In this regard, he was most likely driven by the mediocre – at best – results he had obtained in his battle against the budworm at Westree; he felt under the gun to demonstrate the efficacy of using aerial dusting to control insect infestations in commercial woodlands. Furthermore, unlike the budworm outbreak, which was too extensive for the aerial dusting operation to have had any chance of arresting, there was a bona fide opportunity for the mission in Quebec to succeed. The dominion entomologists who had investigated the infestation reported that it was "more or less confined to the balsam areas near the mouth of the [Manicouagan] river." As a result, Swaine surmised that a precision bombing mission with calcium arsenate just might check the outbreak before it reached uncontrollable proportions.[61]

Propelled by these forces, Swaine took an extraordinary step to push the project ahead. Instead of asking his most senior boss, the Deputy Minister of Agriculture, to file a formal request of the DND's Deputy Minister to carry out the dusting mission in Quebec, Swaine ignored the chain of command and instead directly contacted James Gordon, the head of the Air Board. He did so in early June 1929, stressing in his letter that there was a large and growing looper outbreak in the Manicouagan River valley and explaining that he hoped to dust a small section of it that season. As Swaine put it,

> we are very anxious to carry out an extensive experiment in airplane dusting control so that we can advise the Quebec government and the Industry this season, with confidence, as to the most effective measures for control to be taken in the future. I feel confident that if we can obtain proper weather conditions during the dusting period that the most valuable stands of pulpwood can be protected from this injury by airplane dusting and an experiment such as is proposed covering approximately one square mile should give us the information that we require from the standpoint of both effectiveness and cost of the operation.[62]

He thus asked Gordon to authorize the trimotor Ford aircraft to go to Franquelin, Quebec, immediately after it had completed its mission in Westree to carry out the additional work. In mid-June, Gordon and his Deputy Minister supported granting the request; they were as anxious as Swaine to demonstrate the civilian usefulness of their planes. At the same time, both men asked Swaine's superiors for their concurrence with his request, and for confirmation that Swaine would respect the proper protocol the next time he made a similar requisition.[63]

As improper as Swaine's actions may have been, and although he clearly did not heed the advice he had been proffered,[64] they proved rather inconsequential to the impressive plan that was being hatched. By this time, a host of players from both the private and public sectors were cooperating to organize the project's practical aspects, and Jules-Alexandre Duchastel, the manager of the Quebec Forest Industries Association, was their grand overseer. He had brought together the Quebec Department of Lands and Forests (it agreed to pay for the toxic dust), the Ontario Paper Company and the Anglo-Canadian Pulp and Paper Company (their shared timber limits were threatened by the looper and they agreed to provide the on-site field staff for the effort), and the dominion entomologists and Department of National Defence in order to undertake this fight against this particular forest pest. Once more, the level of cooperation among different levels and branches of government and industry was truly remarkable.[65]

So, too, was the ability of those on the ground and in the air to improvise solutions to the scores of impediments they faced in conducting this endeavour. For instance, the Air Board's officials had initially chosen the village of Franquelin, near the mouth of the Manicouagan River, as the site for their base camp. They were quickly forced to select another location at English Bay, however, after realizing – as one official put it – that Franquelin was totally "unprotected from both wind and sea and is entirely unsatisfactory as a base for the Ford."[66]

In retrospect, moving the project's headquarters must have seemed like the easy part. Flight Lieutenant Ogilvie-Forbes arrived with the aircraft on 5 July, and was ready to begin dusting the next morning. Unfavourable weather initially minimized the available flying time, however, and then mechanical problems began. First, on 10 July the universal joint on the agitator that moved the chemical dust through the hopper sheared, and the plane had to be flown to Rimouski – roughly fifty miles up the St. Lawrence River – to be fixed at a machine shop (Map 1). After Ogilvie-Forbes resumed dusting, the plane's compression struts were seriously damaged when he landed it in rough waters and it was flown to Ottawa for repairs on 18 July. It returned a week later and the sorties resumed. They lasted for a few more days until all the dust had been dropped, signalling the end of the undertaking.[67]

Despite all of the obstacles – or more accurately, because of the ability of those involved in the project to overcome them – the effort in the Manicouagan River

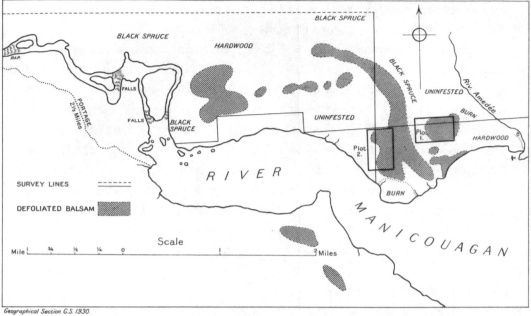

3.5 Map of the hemlock looper damage and the two large plots that were dusted at Manicouagan.
Source: J.M. Swaine, "Airplane Dusting Operations for the Control of Defoliating Insects –
Conducted in Cooperation with the Dominion Air Service in 1929," *Report on Civil Aviation and
Civil Government Air Operations for the Year 1930* (Ottawa: Printer to the King's Most Excellent
Majesty, 1930), Appendix A.

valley proved to be an overwhelming success. This was Swaine's first-ever com-
mercial-scale dusting project to control an insect outbreak that had had a reasonable chance of achieving its aim, and it had hit the bullseye. Over the course of
roughly four weeks in July 1929, Swaine oversaw the dropping of fifteen tons of
calcium arsenate over two large plots, which measured two and a half square miles
in total. The trimotor Ford clocked nearly forty hours of flying time, and Swaine
estimated that it had applied an average dosage of eighteen pounds of toxic dust
per acre. Although he was still encountering difficulties in calculating how many
target insects were killed by the deadly powder, he was confident that the mortality
rate was above 90 per cent, which he described as an "excellent" result. "There is
every reason to expect," Swaine confidently predicted shortly after the project had
wrapped up, "that the looper outbreak has been completely controlled in the lower
Manicouagan."[68]

The project had been a major success in other ways as well. The personnel involved took major strides towards improving how such efforts would be conducted in the future. For example, in terms of marking the target plots and guiding the pilot to them, they determined that it was best to use red and white cloth flags erected on long poles – measuring roughly forty feet – and attached to the trunks of the trees they had topped. They had also learned that the ground crews could direct the pilot most effectively to the test plots by having men in canoes carry poles onto which flags had been attached and situating them in strategic spots on the water. Furthermore, in terms of re-loading the plane with the poison dust, they had come to appreciate that this was done most efficiently by taking a boat filled with the toxin out to the spot on the water where the aircraft landed instead of arranging for the plane to taxi to shore each time it returned to base. Employing this method – which included a chain of men passing the twenty-five-pound bags of calcium arsenate and dumping them by hand into the hopper – had allowed the crew to load 1,200 pounds of dust in only three minutes. Although Swaine hoped to make further improvements to increase the efficiency of such efforts, he calculated that they had conducted this operation at the relatively low cost of six dollars per acre.[69]

Swaine was acutely aware, however, that any success he had experienced on this particular project was definitely not attributable to human efforts alone. Nature itself had played a major role in decreasing the intensity of the looper outbreak along the entire north shore of the St. Lawrence River. No sooner had the drone of the trimotor Ford left the region than the dominion entomologists were reporting that the local looper infestations – even those some distance from Manicouagan – were declining on their own. A few years later they described how the outbreak that had begun in 1928 had "completely subsided," there being "no further traces" of the insect or the damage it had caused: "Trees partially defoliated have made an excellent recovery. The mortality of the severely injured trees is low, probably on account of the scarcity of secondary insects."[70] Perhaps more important was Swaine's acknowledgement that another natural element had contributed in a major way to the dusting project in Quebec producing such a positive result, an outcome that stood in sharp contrast to his experience earlier that summer in Westree. As Swaine explained, "insects which feed in an exposed position on the foliage, such as the hemlock looper and the jack pine sawfly, should be killed readily by dusting." The problem was that "those, like the two spruce budworms, which feed much of the time within nest-like shelter in the foliage, will be found more difficult to kill by this method."[71]

Nevertheless, Swaine was keen to revel in the mounds of dead caterpillars on the test plots in Quebec by publicizing them near and far, and his effort won over

a number of interested observers to his cause. In mid-September 1929, he wrote to
James Gordon, the head of the Air Board, to boast elatedly about the results from
the project against the looper in Quebec. It "was a complete success," he trumpeted,
and "so far as the hemlock is concerned, we feel that the airplane dusting question
has been solved and we have developed an effective and practical method of control."[72] Likewise, he published an article that fall in the *Canadian Woodlands Review*
that reproduced the data from the project. Its outstanding results made clear, he
asserted, that aerial dusting would be a profitable investment for checking small
insect outbreaks and preventing their expansion. The effort also had the desired
impact on its participants. The Quebec Forest Industries Association, for example,
heaped praise on Swaine for the masterful work he had conducted to thwart the
looper's advance in the Manicouagan River valley. Ultimately, subsequent reports
in general would delight in its effectiveness and point to its achievements as indis-
putable proof that Canada's entomologists had turned the corner in their aerial
campaign against forest insects.[73]

While Swaine had been tying up the loose ends from his project against the
looper in Quebec, another aerial dusting campaign of a very different sort was
wrapping up in a fundamentally dissimilar physical and social environment elsewhere in Canada. Over the course of the late 1920s, cottagers and lodge owners
in one of the country's most esteemed recreational retreats had grown gravely
concerned with the appearance of their wilderness haven as it suffered from the
depredations of a native insect whose local population had grown to enormous
proportions. Truly desperate to find a solution, they launched a clamorous campaign to have the government coat their properties from the air with deadly toxins
in order to kill the loathsome forest pest and preserve the aesthetic value of their
landscapes. Swaine and the other dominion forest entomologists were not keen
to spearhead this campaign, but they agreed to oversee and help direct it. And so,
Canada's pioneering aerial campaign against forest insects during the late 1920s
entered a new, somewhat surprising phase, one whose goal was to save nature by
destroying parts of it. To begin telling that segment of the story requires winding
the clock back to the turn of the twentieth century, long before entomologists of
any sort had entertained the notion that aerial dusting would even be possible.

CHAPTER FOUR

"For the sake of this beautiful playground"

KILLING THE HEMLOCK LOOPER IN MUSKOKA, 1927–1929

Canada's aerial war against forest insects between 1927 and 1930 was propelled by the desire to aid more than merely the forest industry in managing its commercial woodlands. Whereas those who converted timber into products such as lumber and newsprint generally saw the forest as a crop to be harvested, environmental preservationists and recreationists revelled in the beauty of the woodlands. For them, the value of the trees lay in keeping them standing for their aesthetic qualities, namely their stately, daunting, towering appearance that was understood to represent nature at its finest. To assist the recreationists in preserving the splendour of their woodlands, Canadian officials shifted their chemical bombing campaign against forest insects literally and figuratively into new territory.

Muskoka, in central Ontario, was the first such site. It was and is arguably one of North America's most exclusive summer vacation retreats. It is prized for its sparkling waterways, stunning outcrops of gneiss rock formations that span a broad palette, and its magnificent forest. In 1927, the hemlock looper began attacking the area's eastern hemlock trees, devouring their needles and turning many of their remaining ones brown. This fundamental alteration of their appearance sent the local cottagers and resort owners into distress and led to their demands that the aerial dusting campaign come to Muskoka.

Recounting how all of this played out demonstrates several themes that are central to this book. The first is the degree to which the local cottagers and resort owners' attitude towards the environment consisted of conflicting and antithetical strands, which clashed on several levels. Although these persons espoused a deep and undying love of nature, for example, their ardour was completely unrealistic.

They conceived of nature as a static entity, whereby each summer season they valued living in the woods only as long as reality closely resembled the same stunning snapshot of the trees, water, and rocks that they held in their collective mind's eye. This conception of nature was completely dissonant, however, with the environment's normal cycle in which the only constant is change. As a result, when their hemlock trees were attacked by a native bug, and thus deviated from their chimerical understanding of the woods, the formerly stately trees lost all value, as did the views from their decks and docks. This necessitated a frantic attempt to return the woods to its previous state, even if it meant employing completely artificial – and highly toxic – means to do so. Ultimately, it entailed manifesting a remarkable inclination to "save" select parts of nature by killing other parts. Attesting to the zealousness of the environmental preservationists in their pursuit of this goal is the lack of data that was collected during these particular dusting projects in Muskoka. The entomologists who oversaw them knew that these efforts were not intended as bona fide experiments to determine, for instance, the minimum dose needed to kill the looper. Instead, their aim was to carpet bomb – with as much deadly poison as they could deliver – a forest pest in an effort to eradicate it.

There was another paradoxical element in the environmental preservationists' attitudes towards the bugs that were feasting on their trees. They consistently couched their case for taking action to combat the pests in the rhetoric of loving nature – an approach that resonated loudly with prevailing social values. Surely, they argued, the sublime landscapes of Muskoka were priceless and were worthy of being preserved at any cost, essentially as a public good. But in truth those who wished to have their hemlock trees dusted with deadly chemicals were motivated strictly by self-interest; this was no benign environmentalism. They were keen to protect their financial investment in nature, and not even nature per se. Occasionally, a cottage owner's appeal for help would belie his true motivations by attaching a finite dollar value to the trees he owned because he knew how much he had paid for their arboreal attributes. In the main, however, the cottagers and lodge owners stuck to the script of rationalizing their drive to realize their own ends by conflating their agenda with environmental stewardship.

Another closely related concept that emerges is the enormous political influence wielded by the urban elite in North America in shaping environmental policy, and the very real dangers inherent in them doing so. Those who sought to "protect" Muskoka from the pests were overwhelmingly "from away." They were not permanent residents in the local area but rather they hailed from the continent's major metropolises, or at least its northeastern ones. They also shared a common attitude towards nature that was remarkably congruous and has been outlined above.

Another element in their frame of mind was their inclination to push obdurately for action to address an issue that they had identified as needing attention. This penchant manifested itself in their steadfast insistence that the government take summary steps to use aircraft to dust their troublesome forest insects even though doing so had, at least in 1928, virtually no chance of succeeding. It also greatly endangered the lives of the pilots flying these missions.

Finally, two other themes figure prominently in this part of the story. The first is the period's unbridled faith in the potential of science to control nature. The vacationers whose trees were inundated with hemlock looper looked to toxic potions, and new delivery mechanisms for the poison, to solve their problems. They had no doubt that they would find the answers for which they were looking in the modern chemists' laboratories and engineers' workshops.[1] In addition, the aerial dusting that occurred in Muskoka during these years once again evinced just how resourceful and courageous Canadians could be when the opportunity to pioneer in the realm of science and technology presented itself. The handful of officials from various branches of government who took on this assignment and saw it through to its conclusion demonstrated an uncanny ability to realize their goals, come what may. This was particularly true of the airmen involved in these efforts for reasons that will soon become obvious.

Practically since Canada's formation in 1867, the Muskoka Lakes have occupied a fabled spot in the country's landscape. They are located in the Parry Sound District in the south-central part of Ontario, and three main and connected lakes – Muskoka, Rosseau, and Joseph – comprise the Muskoka Lakes proper (Map 2). The region itself is found within the Canadian Shield, which gives "Muskoka" its characteristic geomorphology of rugged, undulating terrain, generally shallow soils, and frequent outcrops of red to grey-coloured metamorphic rock. The area is punctuated by a seemingly endless string of lakes, swamps, and waterways, and is part of the Great Lakes–St. Lawrence Forest Region. It is thus covered in a wide range of conifers and deciduous species. On the thinner soils and pockets of deeper, well-drained sands, stands of white and red pine thrive, while hardwoods such as beech, sugar and red maple, and red oak predominate on the scattered and infrequent basins of deeper loams and tills.[2]

After Confederation in 1867, public officials had hoped that Muskoka would become home to Ontario's two most important economic activities – agriculture and lumbering – but they achieved only one of these goals. Having run out of available farmland in the southern sections of the province, the Ontario government initiated a zealous campaign to push agriculture first into the province's central,

and then northernmost, reaches. To stimulate this process, it offered potential settlers virtually free land in these areas in an effort to entice them into undertaking the arduous task of clearing the forest before establishing their homesteads.
Although countless settlers valiantly attempted to realize this aim in Muskoka during the first few decades after Confederation, by the 1880s it was clear that the area
lacked the requisite soils and topography to grow wheat, the intended cash crop.
Around the same time, the local lumbermen also faced significant challenges. The
industry as a whole had formed the backbone of Ontario's industrial economy for
much of the nineteenth century. It had served first as "Great Britain's Woodyard,"
as one author famously put it, and then the dominant supplier of lumber to the
northeastern and mid-western United States. The operators in Muskoka enjoyed
prosperous times for several decades by harvesting the area's fine stands of mature
white and red pine. As the century was closing, however, these stocks were running
short. As a result, some lumbermen began migrating westward and northward to
tap new timber tracts.[3]

These setbacks would not stifle Muskoka's development for long. The dramatic
changes that North American society was undergoing during the late nineteenth
century soon thrust the area into the spotlight for a reason that had nothing to do
with farming or forestry. The urbanization, industrialization, and professionalization
that marked this period had created a relatively affluent class that sought pleasurable
activities to pursue during its leisure time. These same forces had also caused cities
and towns to become generally unhealthy places to live in terms of both physical and
mental well-being, and spawned an ethos that viewed time spent holidaying in "natural" areas as the panacea for what ailed the overworked body and soul. The result was
the creation of a well-to-do group of urban dwellers who had the means and desire to
escape the polluted environments in which they lived and toiled to spend the warm
Canadian summers recharging their batteries in idyllic "wilderness."[4]

This process of conceptualizing Muskoka as *the* spot in which to vacation can
be traced back as early as the 1860s. At that time, recreational fishing and hunting parties began visiting the area, trekking the few existing colonization roads to
reach it. The region was rendered vastly more accessible when a railway was built
in 1875 to Gravenhurst at its southern end, and other lines would soon follow. The
arrival of these urbanites fueled a dramatic growth in the business of the Muskoka
Lakes Navigation Company, which offered regular steamer service to the district's
inviting waterways. Initially it was tough sledding for visitors to the area, as they
stayed in rustic quarters and often tented. Very quickly, however, entrepreneurs
began building luxurious hotels and resorts to accommodate the ever-growing
number of local tourists. The railways were critical to this process, as they greatly

expanded the catch basin from which Muskoka drew its seasonal visitors. Not only did they render cities like Montreal, Detroit, Pittsburgh, Chicago, and New York at most a day's travel away by first-class rail coach, but the railway companies also waged aggressive advertising campaigns to make it appear that Muskoka was but a stone's throw from Manhattan and the Windy City's North Side. By the late 1800s, Muskoka's destiny was sealed as the vacation playground in northeastern North America.[5]

Many of Muskoka's migratory visitors built cottages on waterfront lots on the area's three major lakes, and they typically represented the cream of North American society. Maps were produced to identify who owned which property during this period, and they read like a virtual *Who's Who* of society's upper echelon. President Woodrow Wilson owned an island on Lake Joseph during the 1910s, and Robert L. Borden, Canada's eighth prime minister, was vacationing in Muskoka in July 1914 when fears over imminent war in Europe compelled his private secretary to recall him to the nation's capital.[6]

Muskoka was the destination of choice in eastern North America because it was an easily accessible oasis of natural splendour that represented the antithesis of unhealthy city living; an 1896 publication succinctly labelled it the "land of health and pleasure."[7] Muskoka's boosters waxed eloquently about its pure air and its lakes and rivers' clear and clean waters. "The number, beauty and variety of the lakes in Muskoka is one of the most striking features of its scenery," boasted a late nineteenth-century magazine article about the area, concluding that "Lakes Muskoka, Rosseau and Joseph are her three pearls of nature."[8] Likewise, admirers extolled the virtues of Muskoka's wondrous forests that were renowned for their majestic pine, towering trees that often dwarfed their neighbours in the woods. "[E]very variety of tree life is found to grow here in glorious profusion," the same article on Muskoka proclaimed, and

[t]he forests are immense and imposing in their solitude. The axe of the lumbermen has made slight inroads upon them so they are as primitive and natural as can be imagined ... Hundreds of lakelets are walled in by these vast forests of pine ... On entering [Lake Muskoka] a sense of beauty fills one as the boat steams swiftly into the midst of this lovely sheet of water. Many isles lay around on the bosom of the lake. They and the shores are covered with dense pine and spruce trees. The water is like a polished mirror of steel. The islands are reflected, so are the heavens above. The lake seems as a little sea of glass, brilliant in the sun, skirted with lovely green, evergreen, and the feeling of the place is that of perfect isolation from the world and the rest of mankind. The sun shines on nothing more charming to behold.[9]

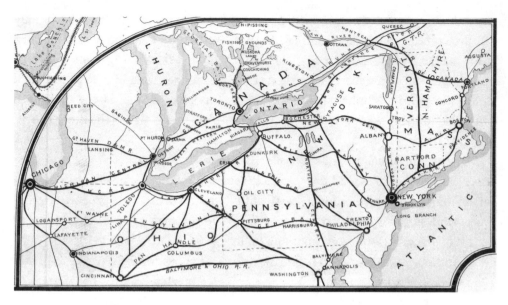

4.1 Tourist map depicting easy rail connections linking the major northeastern American cities and Muskoka, 1904. Source: Detail of Woodward Map of Couchiching and the Lakes of Muskoka, National Map Collection, LAC, e002506308.

Amidst Muskoka's famous white pine are equally impressive eastern hemlock (*Tsuga canadensis*), whose broad trunks and vast expanses often rival those of the awe-inspiring pine. In contrast to western hemlock (*Tsuga heterophylla* (Raf.) Sarg.), which is one of the Pacific Coast's most valued commercial species, contemporary lumbermen generally held the eastern variety in low esteem. Its reputedly knotty and brittle character made it far from desirable for woodworking. The vacationers cared little about hemlock's timber properties, however, as to them it simply represented a beautiful complement to Muskoka's pine. This compelled the area's property owners to nestle their cottages, "bunkies," and boathouses amidst these august trees. The buildings constructed on Chief's and Laurie Islands in Lake Joseph by the turn of the twentieth century compellingly illustrate this inclination.[10]

But during the summer of 1927, Muskoka's hemlock began attracting attention for all the wrong reasons. The area's seasonal residents had started noticing that large swaths of these trees were losing their needles and many of their remaining ones were turning brown, and some even appeared to be dying. The recreationists were distraught by these events because they saw the result as being "ugly." Unsure as to what was causing the problem, they immediately let the Ontario government know that they expected it to take forceful action against this most unwelcome development in their environs.

OUR OWN HOLIDAY PARADISE.

Family Man—No, Dame Fashion, I'm Going for rest and
recreation this summer, and I've found out where to get it !

4.2 Advertisement illustrating Muskoka's irresistible charm, 1888. This image was intended
to demonstrate that, as a recreational retreat, Muskoka ranked above the various locations along
the American Atlantic Coast. Source: *Guide to Muskoka Lakes, Upper Maganetawan & Inside
Channel of the Georgian Bay, the Famous Fishing, Hunting & Pleasure Resorts of Ontario*
(Toronto: Williamson, 1888), 126.

Their entreaties to the public officials were profoundly emotional. For instance, Albert A. Wilks, a cottage owner in the Craigie Lea area at the north end of Lake Joseph, wrote provincial officials in mid-August and relayed a typical view of the situation. "If you could see the fine Hemlocks dying by hundreds on Chief's Island, as I can from our Dock, you would be filled with regret as I am," his appeal for help began. Explaining that he had learned that an insect was causing the damage, Wilks wrote that "all of us around here with cottages and summer homes are hoping you have the matter in hand and will be able to stay the d[i]sease before it jumps to our

4.3 An illustration in a tourist brochure depicting Muskoka's natural beauty, 1888. This print features the towering white pine and hemlock that are seemingly growing out of bare rock, a defining feature of Muskoka's natural cachet. Source: *Guide to Muskoka Lakes, Upper Maganetawan & Inside Channel of the Georgian Bay, the Famous Fishing, Hunting & Pleasure Resorts of Ontario* (Toronto: Williamson, 1888), 8.

shore a few hundred feet distant." Wilks thus asked for advice as to what he should do if the insects crossed the lake and began feasting on his stand of hemlock. He concluded that "it would be a calamity if our trees should be killed by the Hundred's [*sic*] as they are over on Chief's Island."[11]

Another major concern that the recreationists raised was the fear that the changes that were taking place to Muskoka's landscape would impair how the Americans conceived of it and damage its reputation among them. George Freemantle was a local cottage owner and in an impassioned letter he conveyed his concerns about the matter to William H. Finlayson, Ontario's Minister of Lands and Forests. Freemantle described how "a serious blight is affecting a great many Hemlock trees on some of our many beautifully wooded Islands here in this section of Muskoka

4.4 D.C. Mason's recreational property on Chief's Island, Lake Joseph, ca. 1908. This image and Image 4.5 illustrate the majestic white pine and hemlock that defined the rocky outcrops along most of Muskoka's shoreline, and the degree to which recreationists built an array of structures among the stately trees and on the waterways. The photos also highlight what made aerial dusting over forests so difficult and dangerous. The pilot had to fly just above the tree canopy in order to ensure that the poison hit its intended target. The towering spires of different tree species, in this case pine and hemlock, protruded well above the canopy, however, thereby creating obstacles around which the pilot had to manoeuvre his plane. Source: LAC, Frank Micklethwaite, PA-129972.

4.5 Laurie Island, Lake Joseph, ca. 1908. Source: LAC, Frank Micklethwaite, PA-129983.

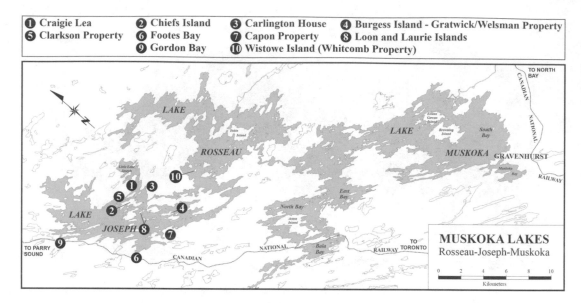

❶ Craigie Lea	❷ Chiefs Island	❸ Carlington House	❹ Burgess Island - Gratwick/Welsman Property
❺ Clarkson Property	❻ Footes Bay	❼ Capon Property	❽ Loon and Laurie Islands
	❾ Gordon Bay	❿ Wistowe Island (Whitcomb Property)	

4.6 Map indicating important reference points on Lakes Joseph and Rosseau, 1927–1929. Map by Léo Larivière. Source: Adapted from AO, RG1–256, Box 1, Hemlock Looper (2), 20 April 1929, D.C. Donaldson to C.R. Mills, to which this map was attached.

Lake." In particular, he stressed that nearly every hemlock tree on Stonewall Island, which was owned by a prominent businessman from Pittsburgh, was now dead. "As you are no doubt aware the islands in this part of our Muskoka Lakes are owned and occupied by wealthy Americans and needless to say they dearly prize the splendid trees particularly the evergreens, the loss of which would be a serious blow." For this reason, he urged Finlayson to investigate the situation.[12]

Proprietors of local businesses were particularly moved to lobby for action, and for good reason. The preservation of the area's natural splendour was literally their bread and butter, and they were determined to protect their manna at all costs. F.J. Ames, owner of the Carlingford House resort a stone's throw from Chief's Island (Image 4.6), expressed similar concerns and predicted dire consequences if the infestation was not brought under control immediately. "A month ago the trees were as they usually are, healthy and vigorous," his letter to the Ontario Forestry Branch began, "but a blight has settled over them and is spreading with a rapidity that is astonishing. The results are nothing short of appalling," he continued, "acres and acres of beautiful trees completely stripped and killed. I went to Chiefs Island, an island of some 300 acres this morning to ascertain the cause and find the place completely over run with caterpillars." Noting that other nearby islands had suffered the same fate, he stressed that "what is so serious is the fact that the blight has reached the mainland opposite Cliff Island." Ames explained that he would not have brought the scourge to the government's attention if it had been limited in scope, "but the extensive and irreparable damage this one is causing coupled with the rapidity of its spread is such as to call for thorough and drastic treatment if a barren waste is not to be the result." He closed by expressing how he hoped, "for the sake of this beautiful playground," that the government would "consent to take necessary action promptly to control the pests responsible for the extensive damage."[13]

Upon receiving word of the problem, Finlayson, Ontario's Minister of Lands and Forests, asked E.J. Zavitz, Ontario's Deputy Minister of Forestry, to investigate it. Zavitz dutifully wrote Peter McEwen, Ontario's District Forester for Parry Sound (within which Muskoka was located), and explained that the minister was receiving letters which described "something that is going wrong with hemlock in that locality." As a result, Zavitz asked McEwen to report on the situation.[14]

McEwen promptly did, and although his observations were deeply worrisome, at least he was able to confirm what the problem was. He openly admitted in a letter to Zavitz on 24 August that "I do not know exactly what is happening to hemlock in Lake Joseph," but he recounted that he had definitely seen evidence of and heard about the same problem "throughout the district." In fact, the dean of the University of Toronto's Faculty of Forestry, Clifton Durant Howe, and a lumberman operating south of Lake Nipissing about a hundred miles north of Muskoka, had described how an infestation of bugs was destroying the area's stands of merchantable hemlock (Map 2). The insects were also having another most vexatious impact. McEwen had learned from the local lumberjacks that the carpet of worms "would bite the men as they crawled over them and they thought them worse than mosquito[e]s." The damage from the bugs was worst, he stated, in pure stands of hemlock along the shores of lakes and rivers, the very spots the cottagers valued most for their beauty. Shortly thereafter, McEwen learned that Belgian-born Dr. J.J. de Gryse, a forest entomologist with the dominion government, had determined that the hemlock looper was the culprit.[15]

After McEwen had personally inspected the trees around Lake Joseph, the specific area about which the letters had expressed concerns, he submitted an update to Zavitz that was filled with despair. He reported that the cottagers were "very concerned," a sentiment with which he empathized because this section of the district represented "the worst example I have seen of damage done by this pest. Many areas of from 10 to 100 acres were seen in which all the hemlock has been completely defoliated and I was assured that these areas are rapidly increasing in size, as the worms are still very active. I do not know if the trees will recover but it is very doubtful that they will." He conceded that "fighting insect plagues is something new to me and I could only tell the people what I had read, and refer them to [government] bulletins. I think it would be impossible to stop it on so large a scale, but would like the opinion of those who have had more experience." Summarizing the situation, McEwen declared that the infestation "appears to be spread over most of the hemlock bearing parts of the district, and to be epidemic in parts, and at the rate at which it has spread from these points it looks very much as if there is a danger of the hemlock suffering the same fate as the tamarack with the sawfly."[16]

For their part, Canada's forest entomologists ruminated over the possibility of using chemicals to deal with the troublesome insects in Muskoka, but they initially dismissed the idea because of what they saw as its practical limitations. Although they had conducted the aerial spraying project in Nova Scotia in June 1927 at the behest of the forest industry, this effort on behalf of cottagers seemed beyond their mandate. As a result, they explained to one of Ontario's forestry officials that ground sprayers were certainly practical for small-scale operations, but the infestation in Muskoka covered a relatively large area. As a result, they saw trying to spray the infestation as a non-starter. Instead, they viewed this particular insect outbreak as offering the dominion entomologists a living laboratory in which to study the looper and determine natural means, such as parasites, that they could employ to combat it.[17]

In contrast, most of the cottage and lodge owners whose properties were relatively small avidly seized the potential solution that ground sprayers offered them. They were prepared to employ any method – even ones employing toxic remedies – to protect the beauty they so deeply valued in Muskoka. As the local government forester put it, a half-dozen cottagers had already purchased forest fire pumps and several thousand feet of hose, and they were going to take combatting the looper into their own hands. "They are willing to go to great trouble and expense in order to protect the trees along shore lines and cottages," he added, "and would be glad for any information as to how to proceed."[18]

Frederick C. Gratwick's attitude was unusually aggressive in this regard. An American businessman and cottage owner on Burgess Island in Lake Joseph (Image 4.6), he had been eager to combat the pest for some time. After realizing his property was infested and learning that spraying poison on the affected trees was the only possible remedy, he hastily wrote a water pump company asking for advice about how to use the machinery to deliver the toxin. He had initially purchased the equipment to protect his property from fire, but now realized that it could be put to better use by dousing the bugs in poison. After the firm reassured him that employing the water pump in this manner would not damage it, he contacted the dominion entomologists seeking more counsel about the toxin they recommended, namely lead arsenate. Gratwick explained that he was anxious to "try to shoot the solution at the trees," and asked them for ideas about which concentration to use and how it ought to be mixed. Recognizing the truly lethal nature of the poison, Gratwick closed by asking, "Would it be injurious to the men putting it on, if so what precautions should be taken?"[19]

Officials with the Ontario government – just like the federal officials in Ottawa – were still at a loss as to what to tell the persons in Muskoka who were beside themselves because of the looper's depredations. For instance, in responding to one anxious resort owner whose property was in the epicentre of the infestation, Edmund Zavitz, Ontario's Deputy Minister of Forestry, despaired that his "Department is unable to cope with leaf-eating insects of this kind where they appear on such a large scale." To another correspondent, Zavitz underscored that, "from our experience with leaf eating insects of this kind it is not physically possible to combat them except in isolated places where [ground] spraying outfits can be used. You will understand that it is very different to carry on spraying operations on large forest areas with forest trees." He reassured the cottagers, however, that the dominion's forest entomologists were on the case and that he was awaiting their advice.[20]

Although the arrival of fall signalled an end to the active insect season across central Canada, the news from Muskoka – and beyond – was not good. The dominion entomologists had reported in the late summer of 1927 on the severe looper infestation at Lake Joseph and also a much larger outbreak in several townships in the northeastern part of the Parry Sound District (Map 2). As a result, not even the brilliant fall colours of central Ontario could distract the recreationists whose hemlock trees were under siege.[21]

They thus kept the matter front and centre before the provincial government throughout the latter part of the year, and each of their entreaties delivered a remarkably consistent message. It reflected their perception that their hemlocks were only beautiful as long as these trees remained as the cottagers believed they ought to appear. Their perspective also accepted that the human manipulation of nature, using whichever means necessary, was an acceptable and even desirable part of enhancing their recreational retreats.

H. Balm's heartfelt appeal in early October epitomized these ideas. In soliciting advice from the provincial government, the Torontonian explained that he owned an eighteen-acre recreational property on Lake Joseph whose fine stand of hemlock was "in danger of attacks of caterpillars." Fearing the worst, he sought advice from the provincial forestry service. "The natural question is," Balm asked without realizing how his question perfectly captured the paradox that defined his perspective on nature, "what can be done to destroy them, is there any spraying device that

I could use with any hope of saving at least more trees on the 5 acres around the cottage?" To buttress his appeal, he outlined how he had been spending significant sums of money planting thousands of young pine seedlings on his grounds, all in the name of improving its future beauty. But, he underscored in reference to his juvenile trees, they were of little value in the present and would remain relatively worthless for decades to come. As he put it, "it would take some years before they are of any appreciable size and in the meantime I don't want to lose my hemlocks."[22]

In insisting that the government take action on this matter, other cottagers took the whole campaign to another level by making particularly forceful appeals. Public officials had an obligation to combat pests that attacked industry's woodlands, they argued, but this duty was exponentially greater when the onslaughts occurred in forests that recreationists cherished for their beauty because such picturesque land- scapes were essentially priceless. Although they often endeavoured to frame their case using the language of quixotic tree-huggers, a few explicitly recognized that the market value of their properties would sink like a lead balloon if the looper were permitted to continue wreaking havoc on their hemlock trees. In other words, they needed the government to take action on their behalf to protect their investments.

D.C. Mason poignantly expressed these sentiments in his correspondence with the government in early 1928. He owned the 240-acre Chief's Island in the heart of the infested area of Lake Joseph (Images 4.4 and 4.6). In a lengthy letter to the Canadian government's Forest Insects Division, Mason described how the problem of the hemlock looper meant "the loss of much commercially valuable timber," but that in the region of Muskoka it meant much more than that, too:

> The value of the Muskoka Lakes is, of course, dependent entirely upon the beauty of the wooded islands and shores, and to realise this it is only necessary to imagine the district without its timber. The timber, therefore, has a value very much greater than the actual commercial value as logs. If this were not the case it would have, no doubt, been cut down before this.
>
> I need not go into the enormous total value of the islands and land surrounding Muskoka Lakes now used for summer resort purposes and the number of settlers and others dependent, principally or entirely, on the summer visitors. The extent of this is obvious to anyone who has any knowledge of the district and is, I think, obviously great enough to warrant consideration of the problem and effective action by the government.[23]

Mason then hit upon the crux of the matter. He explained that his island was nearly half-infested, and how, if the blight spread, he anticipated "losing all the hemlock on the island which will, to a very great extent, destroy its beauty and decrease its value by many thousands of dollars." Emphasizing that many of his neighbours were in "a similar position," he pleaded that "if anything can be done by Government action to protect forest areas from insect destruction, this is a case which most urgently call[s] for it."[24]

Soon enough, a few of those affected by the looper's activities in Muskoka got wind of a potential elixir for their ailment, and their spirits were buoyed by the news. They had learned about the aerial dusting projects that Swaine and his colleagues had conducted in Nova Scotia in mid-1927, and they felt compelled to relay this information to the public officials dealing with the situation in Muskoka. One of them was E.R.C. Clarkson, a high-profile Canadian accountant and a property owner at the north end of Lake Joseph (Image 4.6). In early 1928, he asked the dominion government's entomologists to apply the new aerial dusting technology to deal with the looper in the area around his cottage. If this was not done, Clarkson indicated, "[t]he loss of trees would, of course, make Muskoka a desert waste." T.S. Welsman, who shared Burgess Island in Lake Joseph with Frederick Gratwick (Image 4.6), had heard similar stories and viewed aerial dusting as probably the only way "to avoid the appearance of the island being ruined." Likewise, another cottager predicted that, unless an aerial dusting project was carried out that season, the infestation would "greatly destroy the scenic beauty of this beautiful summer resort."[25]

When Mason heard that airplanes could potentially enshroud the infested trees in poison, he renewed his lobby for action in a way that underscored how the nub of the matter was not environmental but financial. In writing to the dominion entomologists about the situation, he stressed that he actually spent very little time in Muskoka and cared little for the details of what was causing the problem. When asked about the looper's behaviour in the autumn, for example, he could provide no information because he "was only there in the first part of the summer," and because he had not "had the pleasure of meeting the moth and should not know it if [he] saw it." In the same breath he emphasized that money would be no object in combatting it with aircraft and asked only for an "idea as to the cost of dusting by aeroplen [sic]."[26]

The dominion entomologists made it clear from the outset that this was not a project that greatly interested them, but they knew full well that they could ignore the political leverage exercised by Muskoka's seasonal residents only at their peril. As a result, in early 1928 they added Muskoka to the list of locations they were considering for aerial dusting that season. Even if they selected it, however, the decision would come with a number of caveats. Dr. J.J. de Gryse would oversee the dominion forest entomologists' involvement in the effort, and his primary interest in it would be learning about the looper's life cycle and behaviour, not combatting the infestation. Moreover, Swaine, the dominion's chief forest entomologist, stressed that any aerial dusting operation in Muskoka would have a very narrow scope. As he explained to Ontario's Deputy Minister of Forestry, "[w]ith reference to the hemlock looper dusting project our plans include only experimental dusting on a few plots for the purpose of determining the most economical and effective doses." Swaine added that, if the Ontario government decided to undertake "the hemlock looper dusting on a control basis" (i.e., a large-scale program designed to eliminate the outbreak), the Entomological Branch would be "glad to assist in planning and in the direction of the work."[27] Moreover, although de Gryse and Swaine continued to track the infestation in Muskoka in particular and the Parry Sound District in general, they did so only until the fall of 1928, whereupon it disappeared from their radar. By this time, the dominion entomologists had made plain that they wanted very little to do with any aerial dusting that occurred in this famous resort area of Ontario.[28]

Provincial officials, however, could not afford the luxury of limiting their involvement in this way. They continued to be deluged with letters from individuals whose properties were affected by the looper's voracious ways, and the cottagers had ratcheted up their campaign to achieve their ends. In early 1928, the Muskoka Lakes Association, which had been formed in 1894 inter alia "to preserve the safe, healthful and sanitary conditions and scenic beauty of the lakes," had entered the fray.[29] Not only had it organized a concerted letter-writing campaign to lobby the Ontario government to take action against the looper but it had also struck a special committee to take up the matter of the "hemlock blight" with provincial officials. It was headed by Sir Thomas White, a venerable Tory who had served as Canada's Minister of Finance during the First World War and was a director of some of the country's leading financial and industrial concerns by the time of the hemlock outbreak.[30]

White proved remarkably adept at compelling the Ontario government to combat the looper, a mission that was undoubtedly facilitated by the common political stripe he shared with the ruling provincial party. White immediately arranged a private meeting to discuss the matter with William Finlayson, Ontario's Minister of Lands and Forests. The Muskoka Lakes Association informed its constituents a short while later that, as a result of this tête-à-tête, Finlayson had agreed to purchase "a special aeroplane to be used to dust the areas affected and wipe out the pest that has threatened the Hemlock trees in Muskoka." To pay for the aircraft, Finlayson quietly slipped a requisition for $20,000 into his government's supplementary estimates in mid-March 1928, and the Ontario Forestry Branch then began planning a dusting operation on a control basis against the looper.[31]

Ontario now had to go shopping for an appropriate aircraft, and for guidance it turned to W. Roy Maxwell, the Director of the Ontario Provincial Air Service (OPAS).[32] Predictably, Maxwell contacted the Canadian Department of National Defence (DND) for advice since it had provided the plane and pilot for the inaugural forest dusting experiment the previous summer in Nova Scotia. Maxwell knew that the DND had used a Keystone Puffer on the project because it had proven highly effective in dusting cotton fields in the American South and wheat fields in Manitoba. He thus asked for detailed information about the Puffer's performance during the forest dusting work in the east.[33]

When the reports began arriving, however, they were replete with warnings. The one from C.L. Bath, the DND's pilot who had flown the Puffer during the forest dusting sorties in 1927 and whose account was cited in the previous chapter, unequivocally declared that the aircraft was completely unsuited to forest dusting. Bath's commanding officer the previous season and the Director of Technical Services for the Air Board, Wing Commander Ernest Stedman, wholeheartedly concurred with Bath's conclusion.[34] Likewise, James Gordon, the Air Board's director, echoed this message a short while later and was emphatic that "further work of this kind [should] be undertaken with a plane having a greater factor of safety against engine failure, preferably a tri-motor, so that in the event of one engine failing, it would be possible to fly to a safe landing place."[35]

In the face of this news, Maxwell had to change gears, but this was easier said than done. After investigating alternative aircraft, he concluded that it would be best to purchase a de Havilland 61 (DH61) for the dusting project in Muskoka. Although it had only a single motor, its engine was much more powerful and its payload much larger than the Puffer's. The official historian of the OPAS describes it as having been "the first big 'modern' aircraft" the OPAS acquired.[36] By late April, the Ontario government had ordered its DH61 from the manufacturer in England, and although it was scheduled to arrive within a month or so, by early June there was still no sign of it. This was problematic for a couple of reasons. Firstly, the delay left relatively little time to fit the plane with a custom-made hopper. Even more importantly, however, the dusting had to begin by mid- or late June, the period during which the larvae were small enough to be most vulnerable to the dust's toxic effects.[37]

This second factor had been the reason why the dominion entomologists had pressed the Ontario government to begin dusting in Muskoka – in the worst-case scenario – around the time of summer solstice. Swaine and de Gryse believed that Ontario could launch its campaign against the looper by this relatively late date because they had assumed that the province would carry it out using the new DH61, whose relatively large payload capacity would allow it to conduct the dusting over a much shorter period of time than the Puffer. As a result, de Gryse, who advised Edmund Zavitz, Ontario's Deputy Minister of Forestry, on the project in Muskoka, had arrived on scene by 20 June and had immediately begun providing updates on the status of the looper's development that season. Three days later he telegraphed Zavitz with a brief report: "Larvae in first stage. Stop. Very little feeding. Stop. Infestation general work should begin as soon as possible. Stop. Area very extensive. Stop. Will be difficult for one plane." But the DH61 was nowhere to be found one week later, when de Gryse sent Zavitz a letter from Footes Bay (now Foot's Bay), the local base for the dusting operation (Map 2 and Image 4.6). After updating Zavitz on the infestation, de Gryse emphasized that "[t]he majority of the larvae are now in the second stage. The sooner dusting can begin, the better will be the results."[38]

The cottagers in Muskoka were probably oblivious to this entomological consideration, but they were certain of one thing. The Ontario government had promised them that it would dust that season, but the cottagers had not yet heard the aircraft's drone. In fact, they had begun pestering Peter McEwen, the local District Forester for Ontario's Department of Lands and Forests (DLF), with inquiries about the dusting work back in early June. Frederick Gratwick was particularly insistent in this regard. He had sought advice about rigging up his own spraying apparatus in August 1927 to fight the looper on Burgess Island, which he owned. Determined to protect his property investment and that of his neighbours, he requested that the government dust the infested hemlock on their island. In response, Zavitz assured him that the DLF was preparing to carry out dusting operations in the infested areas very soon, and that his requests had been duly noted.[39]

Gratwick grew deeply worried when still no action had been taken by early July, and so he contacted the DLF again to express his heightened concerns. He informed provincial officials that "we are very much disappointed that the aeroplane for dusting the hemlock loopers has not yet arrived. Will you kindly let me know when you expect to have the plane there?" He reminded the government that the looper had been "very destructive" the previous year, and that the early indicators pointed to even worse damage occurring that year. "If this pest is not checked," Gratwick postulated, "it is going to ruin that section as the hemlock is one of the best trees that we have. A very large area of the shore line is at present well wooded with hemlock, and if these remaining trees are killed as hundreds of them were a year ago, it is going to ruin the region. Will you be kind enough to give this matter your urgent attention and advise me … when we can expect assistance from the Government?"[40]

One cottager expressed more annoyance than anyone with the delay. F.J. Capon owned a cottage on Lake Joseph's south shore (Image 4.6), and it soon became clear that he had somehow managed to jump the queue of those who wished to have their properties dusted. In a curt telegram to Zavitz on 12 July, Capon asked tersely, "[l]ooper gaining fast why delay with plane[?]"[41] While Zavitz would soon provide the reason, it was not as if the Ontario government had been sitting idly by. For a few months, its officials had been laying the groundwork for the aerial dusting project in Muskoka. This entailed ordering sixteen tons of calcium arsenate and arranging the lodging for the personnel involved in the project. All that was needed was for the DH61 to reach Canadian soil, but when it finally did, the news was not all good. Zavitz wrote to Capon in mid-July with a somber update: "[P]lane for dusting damaged in shipment from England. Stop. Repairs being rushed probably available early next week."[42]

This proved to be wishful thinking on Zavitz's part. The disabled DH61 had been sent for repairs to the OPAS's headquarters in Sault Ste. Marie, about 250 miles northeast of Muskoka (Map 1). On 12 July, Zavitz sent an urgent telegram to Sault Ste. Marie that asked when the aircraft would be ready and added that he was "very anxious as season for dusting is nearly over." On the same day, he learned that the plane had suffered substantial damage, and it would be some time before Canadian aeronautical officials certified it as being airworthy.[43]

With the cottagers' anxiety-ridden letters continuing to arrive on Zavitz's desk – most pointedly asking about "the aeroplane which is overdue"[44] – he was pressed to devise a quick solution, and he did, albeit one that entailed significant risk. On 13 July, he sent a telegram that desperately asked the Director of the Air Board if he would loan the OFB a Keystone Puffer for special dusting operations in Muskoka. "If so," Zavitz continued, "have your pilot deliver machine at Footes Bay on Lake Joseph Saturday fourteenth instant if possible. Stop. Our pilot will be there to take over Saturday night or Sunday morning. Advise."[45] The officials in Ottawa agreed to comply with this request, but only under two significant conditions. Because they considered the Puffer so ill-suited to forest dusting work, they would supply only the plane, and not a pilot to fly it. In addition, the RCAF tellingly demanded that the Ontario government "make good any damage incurred" by the Puffer during the course of the operation.[46]

Zavitz hastily accepted the second caveat, and he eagerly endeavoured to overcome the challenge posed by the first. He sent Maxwell a telegram that asked if the OPAS could supply a pilot to fly the Keystone Puffer until their own plane was ready: "Ottawa will loan us puffer and fly it to footes bay [sic] but will not supply pilot for operations. Stop. Dusting must start at once." Maxwell wired back that he was "instructing pilot [E.J.] Cooper [in] Sioux Lookout [to] report [to] Footes Bay immediately." Zavitz realized that it would take Cooper at least a couple of days to fly down from his distant field station in Sioux Lookout (about 750 miles away in northwestern Ontario), which was time Zavitz felt he could simply not afford. As a result, he pressed Maxwell to find a more immediate solution, which Maxwell did. While Cooper would still report to Footes Bay, Maxwell directed another pilot, G.R. Hicks, who was stationed in Sault Ste. Marie (only about 250 miles from Muskoka), to head there at once. Significantly, when Maxwell ordered his two pilots – Cooper and Hicks – to report to Muskoka, he provided them with no information about their mission's obvious danger. His directions simply read, "[i]mportant dusting operation Muskoka. Air Board will loan us Keystone Puffer but we must supply pilot."[47]

Maxwell's decision to remain silent on the issue mattered not. Hicks, who was Ontario's first pilot on the scene at Footes Bay, soon learned of the unsettling reports about the Puffer. Flying Officer Bath had been resoundingly critical of the Puffer after he had flown it during the dusting experiment in Nova Scotia in 1927, and the Air Board had chosen him to fly it from Ottawa to Footes Bay on 16 July. Upon his arrival, Bath had communicated to Hicks his grave concerns about the plane's suitability for forest dusting.[48]

Hicks immediately dispatched a lengthy telegram to Maxwell that conveyed this disappointing message, and another one, too. Hicks began by telling Maxwell that de Gryse, the dominion entomologist on the project, had stated that it was "futile to dust after 28 July and there is one month['s] work if flying carried out daily. Stop. Considers it useless to send machine as only few strips can be done out of 1500 acres." Hicks then described how "Pilot Bath informed him that Puffer dangerous for work here and personally would refuse to undertake it. Stop. Essential that machine with characteristics of … [the DH61] … be untilized [sic] for this hazardous flying. Stop. Success of operations depend[s] on machine and Pilot trained in dusting operations. Stop. Does Cooper know anything about it. Stop. I do not. Stop. May be possible to complete fifty percent of work if big machine [i.e., the DH61] sent now."[49]

Maxwell endeavoured to assuage his pilot's fears, but his logic was as faulty as the Puffer was dangerous. "You will be replaced by Cooper as early as possible," Maxwell reassured Hicks, although he informed Hicks that Hicks's desire to carry out the dusting with the DH61 instead of the Puffer would not be possible because the hopper could not be installed quickly enough to permit the DH61 to be used on dusting operations that season. He also asked that, in the future, "further complaints and information as contained in your wires be forwarded to [him] via letter."[50]

Hicks obliged by penning a message to Maxwell that laid out the numerous additional problems he was facing. He stressed that his previous communications "were not to be read as complaints," but rather were conveying information he thought "may be required in the near future if the threats of the personnel operating here materialise." Hicks continued:

The Chief Entomologist [i.e., de Gryse] who was hired for this work informed me that he was told that I was proficient in dusting and had in fact been employed in this work for the last few years. This was told me after I had told him that I had never done it in my life, but I would do my best to assist him. I could not retract my statement once it had been made [and] consequently my inability combined with other annoyances to which he has been put here, compelled him to make a strong protest to the powers when his work here is completed.

I am told that a number of influential men residing in this district were promised that the whole of it would be dusted this year as the trees are being destroyed [and] in consequence the value of the district is depreciating considerably. One man, a millionaire, has threatened all sorts of things if the area adjoining his house is not dusted. The entomologist states it is futile to dust the trees after the 28th July because the bugs cease to eat by this date. Over 1500 acres are to be dusted [and] with the Puffer this area will take 30 days, flying every day. The slightest breeze destroys the operation. So you can see that ideal weather conditions are necessary if success is to be attained. Thus two months is required. The entomologist was promised the machine by July 1st [and] because he was led to believe that the DH61 with its carrying capacity of one tonne would be ready, he chose this date, but had he thought that the Puffer would be used he would have asked for it by June 1st, or failing this would not have accepted the job.[51]

Nevertheless, Hicks dutifully followed his orders and in the process faced the inherent dangers head on. After learning of the Puffer's deficiencies on the evening of 16 July, he decided to take the aircraft for a test flight the next morning. He reported to Maxwell that his forty-minute trial with the maligned aircraft "confirmed all the adverse reports [he] had previously heard regarding its performance." All the same, Hicks had ventured forth because he "felt that a contrary decision would discredit the Provincial Air Service as air news travels quickly and is often misconstrued."[52]

And venture forth he did. After his very disappointing test flight on 17 July, Hicks was ready to begin dusting the next day. Rain washed out any hope of doing so until late afternoon, but when the weather finally cleared, Hicks flew two sorties during which he dropped a total of 600 pounds of calcium arsenate. His target was a hemlock stand on the mainland in the southern part of Lake Joseph, and "[t]he greater part of the area dusted belongs to Dr. Capon." Hicks had flown too high over the trees, however, thereby allowing the strong wind to blow most of the poison out into the lake. The next day he dusted the same area, but this time the wind was very light and he flew not directly over the trees but along the shore as close to the trees as possible. The onshore winds thus gently drifted the dust into the target trees, covering them in a thin white coating of poison. The following day, Hicks used the same technique to dust two islands – Loon and Laurie (Images 4.5 and 4.6) – with 600 pounds of chemical with equally satisfying results.[53]

Hicks surmounted a succession of hurdles in performing this work. In a letter to Maxwell, he vividly described the effect of the dust on his eyes. "As soon as the hopper is opened," he lamented, "the dust pours in from it to the cockpit, and the famous 'Luxor' goggles are useless to keep it from my eyes. My word how it stings [and] it gives me a headache every day. It had the same effect on Flying Officer Bath when he commenced dusting, but he has since got used to it. Holy Mackerel what a rotten machine the Puffer is. The floats are absolutely wrong … She stalls at about 75 M.P.H. [and] loses speed horribly on turns."[54]

Just when it appeared things could not get worse for Hicks, they did. He decided to tolerate these difficulties until after he had completed his dusting mission on 20 July. At that time, Flying Officer Bath, who was still at Footes Bay, passed along more discomfiting information to Hicks. Although Bruce West, the OPAS's official historian, describes Hicks's typical reports as having been "quite business-like and matter-of-fact," Hicks's recent experiences pushed him over the edge. As a result, West notes in reference to Hicks that "it seemed that all the English reserve he could muster could not completely conceal the understones [sic] of sheer horror which crept into his report." The reason was that Hicks had learned from Bath that "the Puffer machine is to be disposed of by the Air Board" after Bath and Squadron Leader Grandy had "strongly recommended this, because in the opinion of both, especially the latter, it is 'suicidal to fly the machine on dusting operations.'" This was why the DND had "most willingly loaned it to this Service [i.e., the Ontario government] on condition that the R. C. A. F. pilot did not participate in dusting operations."[55] Hicks was naturally upset by this revelation, but he was acutely annoyed because, as he put it, "I was not given these facts until the 20th, otherwise

4.7 The Keystone "Puffer" that was initially used to dust Canada's forests, ca. 1928. This photograph vividly captures why it was so dangerous to use the Puffer for dusting Canada's forests. Note the stains of poison dust on the lower fuselage and the position of the cockpit. When the dust was released from the hopper near the plane's nose, it created a toxic cloud that both blinded and poisoned the pilot. Source: National Aviation Museum (Ottawa), Negative No. 6103.

my decision to fly would have been the reverse to what it actually was, because I feel that my action in carrying on has caused ill feeling between the Services [i.e., the OPAS and the RCAF]." Hicks's frame of mind was undoubtedly poisoned further by the events that occurred during his final flight. After opening the hopper door, the dust had again enshrouded him. This time, he revealed in a letter to Maxwell, it "completely blinded me until I got more or less used to it. On the 20th the air was humid and the dust which fell on my goggles formed a film of paste and my efforts to rub it off resulted in my hitting two trees with the pontoons. Fortunately with no damage to the machine."[56]

While Hicks had survived his ordeal with the Puffer, it was only beginning for another of the OPAS's pilots. Maxwell ordered Hicks to hand over the flight controls to Pilot Cooper, who had just arrived from Sioux Lookout. Eager to find out for himself if the horror stories about the Puffer were true, Cooper took off first thing the next morning on a short dusting test run over Chief's Island (Images 4.4 and 4.6). After another flight with the Puffer, Cooper refused to fly the plane again. "Upon landing," an observer noted, "he stated that he considered the machine in used [*sic*] (the Keystone Puffer) as unsafe and that he did not know whether or not he would continue the work begun by Hicks." Cooper actually had no choice in the matter. When he had set the plane down after his final flight, "it was crippled in landing."[57] Incidentally, the Air Board's annual report for 1928 did not mention one word about the dusting project in Muskoka that season even though one of its planes had been used in the effort, an omission that strongly suggests it was keen to avoid being associated with this extraordinarily hazardous undertaking.[58]

By this time, there were other equally compelling reasons for cancelling the project just days after it had begun. de Gryse outlined them to the Ontario government on 22 July. "I regret exceedingly to feel obliged to take such a course," he explained, and "[t]he main thing to be regretted, however, is the failure of the provincial dusting plane specially purchased, as I understand it, for this work. When I saw Mr. Zavitz [Ontario's Deputy Minister of Forestry] the last time, he told me that he was sure the plane would be ready for the first week in July. I based myself on this in advising him to order the dust."[59] As a result, the project used only 2,500 pounds of calcium arsenate dust out of the total 32,000 pounds that had been ordered, and de Gryse assured Zavitz that the dust would keep over the winter and it could be applied the following June. In summing up the situation, he philosophized that "[t]o try to shoot a pile of dust on the trees, at this time, is simply laying ourselves open to ridicule."[60]

Notwithstanding all this travail, from an operational perspective the dusting in Muskoka in 1928 produced a few important results. Although it treated only a relatively tiny patch of land, the pilot had engineered a new, far more practical method of applying the dust that no longer required complete calm as long as the targeted stand of trees was close to the shoreline. Pilot Hicks had realized that he could fly parallel to the water's edge and have the breeze waft the dust into the infested hemlock. Not only was this safer than maneuvering the plane over the forest's uneven crown, but it applied the dust much more evenly to the trees in the littoral areas. As de Gryse described it, the trees were left "practically white with dust and a very good dose of poison was found on all growth as far as the hemlocks extended." In addition, the dominion's forest entomologists conducted small-scale experiments with larvae on poisoned foliage. These trials convinced them that the small larvae were readily killed by the calcium arsenate dust, whereas the larger ones could withstand "a much heavier dose than that which it was possible or practical to apply by airplane." Ultimately, de Gryse concluded that "airplane dusting for the hemlock looper is not only feasible but eminently practical, if done at the proper time" (i.e., when the larvae were very small).[61]

For their part, the cottagers were deeply dismayed by word that the dusting effort had ground to an unexpected, premature halt. Hicks had reported immediately after the cessation of the project that "[m]any verbal battles took place between the land owners in the District who each considered that their property required dusting first. Mr. DeGryce [sic] endeavoured to pacify the wrathful owners but I fear to no avail, because apparently they were promised that their property would be dusted this year and consequently would accept no excuse." Hicks continued, "Various threats were made, such as taking their money out of the district, taking up the matter with the Minister and last but not least appealing personally to the Prime Minister. Whether these threats materialize remains to be seen."[62]

Whether they did is unclear, but the cottagers continued to dispatch irksome letters to the provincial officials. It was understandable that this deluge peaked during August, the height of the "season" in Muskoka, and most of these pleas asked for the government to treat a specific area – namely the letter-writer's property – with chemical dust. These requests also conveyed the sense of hysteria and siege mentality that had enveloped the cottagers' collective mindset. For example, E.A. Apple, Secretary of the Gordon Bay Association (Image 4.6), wrote to the Ontario government on behalf of his Association. He explained that he had heard that the authorities had a "scow down the lakes, spraying looper areas," and he declared that it was "imperative" for the government to move this operation to the Gordon Bay area. To buttress his case, he postulated that, if the looper was "not gotten under control, it will also tend to kill off the deer, as the deer in winter feed mainly on hemlock browse."[63]

A few of the correspondents' letters were particularly noteworthy for their willingness to distill the battle against the looper down to a matter of dollars and cents. O.W. Clapperton owned Camps Caribou and Sagamesing well north of Muskoka, and he was aghast at the scene that was unfolding in front of his eyes. He informed the Ontario government that he had paid $16,000 for one of his resorts, which was located "in a thicket of fine Virgain [sic] Hemlock timber" that was now "badly infested with this loop worm." As a result, he pleaded with the public officials to take action – or help him do so – to protect his investment.[64]

These concerns were expressed even more forcefully by an American, Edgar B. Whitcomb, owner of Wistowe Island on Lake Rosseau (Image 4.6), in several impassioned letters. He addressed the first to Ontario's Deputy Minister of Forestry in late August 1928:

We are in great trouble here on our Island because a blight has suddenly attacked one stand of hemlocks which was very beautiful and killed all the trees within one month. There are two other stands, one surrounding the house and if these should go this fall or next year, we should no longer possess a place of which we have been inordinately fond and very proud. It would be almost like losing a dear friend. I would far rather suffer a loss of buildings or boats by fire as that would be only money and replaceable. I am very much rattled and am naturally turning to you for help or advi[c]e. Have seen the same thing happening elsewhere last year and this but our place has been spared as have still some of the other islands. I have no doubt you have been aware of the trouble and are at work on it but I do not desire to await what now seems inevitable and to endure further loss without a struggle and would very much appreciate it if the department has any advi[c]e to give or assistance to offer. I am willing to incur any expense within reason and work hard to save the day.[65]

Whitcomb's missives to the government also fittingly captured how the recreationists believed humans could and should take steps to keep their retreats in the forms they considered most aesthetically pleasing; having them fine-tuned by human hands was a perfectly normal and even desirable part of appreciating these sights. For in Whitcomb's eyes, public officials could and should exercise significant agency under the circumstances that prevailed in Muskoka. As he put it, "I know of nothing that has come up during the fifteen years of our summer residence at Muskoka that has been so serious as this trouble may be. Low water was one annoyance I recall but it was curable and, when cured, left no trace. Other things were much of the same class. In this case, if the hemlocks are destroyed, many of the islands will be partially ruined in the matter of being attractive and may be abandoned by their owners for that reason. I have in mind, when saying this, the effect of severe forest fires." As a result, Whitcomb asked for the Ontario government to assure him "that the subject is before the Department and will have effective action in the near future ... I have not been so worried over anything connected with Wistowe Island during the ensuing winters as I am about this matter this winter."[66]

For good reason, the Ontario government would soon be able to make just such a vow. Although the dusting in Muskoka that season was both disappointingly brief and dangerous, the dust had proven deadly against its target insect. de Gryse's reports had confirmed, in his mind at least, that the hemlock looper could be controlled by aerial dusting with calcium arsenate. This news had imbued the Ontario government with complete confidence that it had found the solution to problems involving this forest pest. As a result, in answering the concerns raised in mid-August 1928 by Joseph Penrose, a cottage owner whose property was being, in his words, "utterly ruined with this pest,"[67] Charles R. Mills, Ontario's acting Deputy Minister of Forestry, offered strident assurances. He triumphantly pronounced that "[t]he only method by which the Hemlock Looper can be controlled is through dusting from Air Craft." Mills also pledged that the provincial government "intended to do everything possible to control the infestation during the early part of next season when the caterpillars are small and can be killed by dusting."[68]

But to fulfil this commitment the Ontario government would first have to overcome one potentially insurmountable hurdle. W.R. Maxwell, the Director of the OPAS, whose pilots had flown the missions in Muskoka in 1928, outlined the roadblock in a letter to Mills in late August of that year. He pointed out that the "dangers" connected with flying the Puffer over forests had caused all the pilots who had done this work to decline the invitation to do so ever again. Maxwell explained that pilots Hicks and Hector Ptolemy, both from the OPAS, "advise that they do not desire to pilot machines under dusting programmes. Of course if formal demand or request is made they intimate that they are not afraid to do so, but are both quite frank in stating that they would be pleased if another pilot selection could be made." Similarly, Maxwell reported that Flying Officer Bath, from the RCAF, was "also very much against carrying out further flights of a similar nature."[69]

So over the course of late 1928 and into early 1929, personnel with the Ontario Forestry Branch worked at removing this particular roadblock from their path. They sought advice from Bert Coad, the American guru of aerial dusting, and they worked on constructing a hopper that would work with their DH61, which they had now repaired. This allowed them to test during the autumn the plane's suitability for aerial dusting, trials whose success overcame the pilots' aversion to flying these missions. A few days before spring arrived in 1929, the OPAS declared its aircraft battle ready for bombing the looper in Muskoka with toxins.[70]

This was a most welcome update for the Ontario Forestry Branch, for the scope of the problem with which it was grappling had grown exponentially. Ever since it had abruptly terminated its dusting project in July 1928, its officials had been instructed to compile a list of the recreationists who had complained about the looper and requested that the government treat their properties. While most of the concerns had been expressed by those on Lake Joseph, a small area of which had been treated the previous season, a significant number had come from cottagers and lodge owners on Muskoka's other two major lakes, and seasonal residents farther north and west. Their properties were located on Lakes Muskoka and Rosseau, a few islands and lakes near Parry Sound in Georgian Bay, and Ahmic Lake (roughly halfway between Lakes Muskoka and Nipissing). The government's foresters investigated and mapped the local looper infestations during the summer of 1928, and their findings confirmed the cottagers' reports (Map 2).[71]

With the Ontario government having pledged to each property owner who had complained of the looper menace that it would treat their infested trees, it was now essentially operating a private aerial dusting service at the beck and call of some of the province's – and northeastern North America's – most well-heeled recreationists. Consequently, even though the Ontario government had nearly 15,000 pounds of calcium arsenate dust left over from its aborted project in Muskoka in 1928, it was compelled to place a new order for the same volume of chemical for its aerial campaign in 1929 because the demands upon it had grown so large. It would dispense most of the dust in Muskoka, but it sent a large volume of the toxin farther north to the Parry Sound District for dusting there as well. Upon hearing of this news, the media in Toronto were fixated with the notion of what awaited the forest pests in Ontario's renowned cottage country. "Timber Insects Will Dine on $6,000 Meal of Poison," screamed one of the more extreme headlines in early 1929, a message that confirmed that the Ontario government would be footing the entire bill for the dusting operation.[72]

As promised, the Ontario government delivered the toxic feast right on time that season, but only because those directly involved demonstrated remarkable ingenuity and creativity in overcoming the problems they encountered. In Muskoka, over the course of 19 June to 10 July, provincial officials coordinated the dropping of just over 35,000 pounds of calcium arsenate – at a rate of roughly 30 pounds per acre – over an approximate total of 1,200 acres of looper-infested forest. The OPAS's pilot, Hector Ptolemy, flew the missions using the DH61, and his adept aerial skills drew praise from all concerned.[73] But the effort only went forth after the support crews were able to solve a number of mechanical problems. For example, the first sortie Ptolemy flew revealed that the tube that released the dust from the hopper was constricted, and so the plane was dispatched to the OPAS's base in Sault Ste. Marie for repairs. Upon returning it performed well, but then the lever that the pilot pulled to release the dust broke. Fortunately, one resourceful mechanic visited the Canadian Pacific Railway's industrial shop in nearby Mactier to build a new mechanism, which was ready post-haste and was actually better than the original version. Thereafter, all the persons who had requested that their properties be dusted had their wishes met, and as a result some of Muskoka's most famous islands – Eagle, Loon, Reef, and Wistowe – were cloaked in the deadly powder. The aircraft was transferred to Georgian Bay on 10 July, and for two days it continued its work there. Over 11–12 July it was able to treat just over 250 acres of land on Otter Lake (just south of the town of Parry Sound) and a few islands in Georgian

Bay with nearly 5,000 pounds of dust (Map 2).[74] But then the plane and pilot were called away on an emergency mission, thereby putting an abrupt end to the project. In total, the effort dusted cottagers' properties and adjacent waterways in nearly twenty tons of calcium arsenate.[75]

The entomologists must have been gratified by what they observed during their follow-up investigations. They had grown concerned about the potential efficacy of their work because the period during which they had treated the area had been punctuated by irregular winds and heavy rains. Moreover, the dusting had been repeatedly delayed because they had been forced to address the problem of the DH61's engine overheating by letting it cool for nearly one hour between each sortie. Furthermore, there was significant evidence that several pockets of the infestation had begun dying back on their own. Nevertheless, the toxin proved stunningly effective. The entomologists' tallies of dead and dying loopers confirmed that aerial dusting against epidemics of this particular insect could succeed in controlling them. As de Gryse informed the head of the Ontario Forestry Branch, "the Muskoka Lakes dusting may be considered a real success."[76]

About one decade later Arthur Gibson, the Dominion Entomologist, reflected upon the successful nature of this project, and in doing so his words brilliantly captured the central belief that had both fueled this particular pioneering effort in aerial dusting and was indeed pervasive across North America; no price was too high to pay for protecting trees that society considered beautiful. In an address at an international entomology conference in Germany in 1938, Gibson recounted how Muskoka had "developed into one of the most popular summer resorts in Canada, and the hemlocks and other trees have, therefore, a high aesthetic value." He explained that the hemlock looper's local population had exploded in the mid- to late 1920s, however, and that the insect had begun devastating the local woods. As a result, "an extensive airplane dusting project was undertaken by the Ontario government, supervised and directed by officers of the Entomological Branch." Then came the crucial part of his message, specifically that "this method of control proved effective, and although costly, the value of the trees justified the expense."[77]

There was a remarkable aspect to Gibson's comments. Although he spoke to only one of the main issues that lay at the core of the dusting projects in Muskoka – he said nothing about the incongruity of the recreationists' views of nature, for example – the forces that had driven them would propel several other similar efforts thousands of miles away. Out in British Columbia in the late 1920s, for instance, officials in the Lower Mainland were dealing with several insect infestations that were destroying some of the residents' favourite woodlands. Paradoxically, none of these people perceived of their own intrusions into the woods – in the form of operating

a luxurious lodge or manicuring the trees in a park or damming a river or drain-
ing a lake – as sullying the "untouched" nature of their forests. Looking out at the
magnificent trees, it mattered not what instrument was needed to maintain the
observers' staid and subjective conception of them. It mattered only that one could
be found. And so, over the course of 1929–30, recreationists and city boosters alike
would call for – demand, really – the application of deadly chemicals in an effort
to preserve their fanciful view of the woods that graced Canada's westernmost
province.

CHAPTER FIVE

"You cannot control an infestation such as this with toys"

POISONING FOREST PESTS IN BRITISH COLUMBIA, 1913–1929

The themes that pervaded the aerial dusting projects in Muskoka also defined those that occurred in British Columbia over the course of 1929–30, but as with so many aspects of Canada's history, they occurred with a slightly different twist in BC. The effort that was undertaken in that province in 1929 was aimed at protecting trees around the Wigwam Inn, and once again environmental preservationists took a paradoxical approach to the woods. They openly declared that they worshipped nature for its pristine and aesthetic qualities and yet were eager to kill some parts of it in order to save others. They were also blind to the degree to which they altered the forest because they saw their work as improving it. Most of all, those who pushed for dusting the Wigwam Inn in 1929 did so on the grounds that the work was of inestimable value because it aimed at preserving scenic beauty. Nevertheless, on a few rare occasions they admitted that the true animus behind their actions was the desire to guard an expensive asset.

There are other similarities between the dusting project in BC in 1929 and those that occurred in Nova Scotia, Ontario, and Quebec. For example, political factors were crucial on several different levels. Powerful figures from various realms dictated where and when this aerial dusting occurred, and cooperation among different branches and levels of government was essential to realizing it. The fact that BC has often acted as a distinct entity in Canada also determined that all the dusting projects that occurred in the province were discrete from those that occurred back east. These circumstances compelled those involved in the operations on the Pacific Coast to reinvent the wheel in many respects. In doing so, they demonstrated the same trailblazing skills and spirit that characterized the projects that had occurred in eastern and central Canada. Finally, nature steered the effort at the Wigwam Inn in its own special way, and repeatedly reminded those who were involved in the project that they alone would control neither the bugs nor the trees.

Among Canada's provinces, British Columbia was arguably the country's debutante during the early twentieth century. Having been created in 1870 as a relatively tiny, isolated colony made up of barely 10,000 non-Indigenous residents, a few decades later it was riding a veritable tidal wave of growth. Its population was burgeoning thanks largely to the development of a string of industries based upon extracting BC's profusion of natural resources. Foremost among these enterprises was the lumber business, which was capitalizing on the province's truly prodigious supply of virtually untapped timber, much of it highly valuable. This industry was given an enormous boost when the Panama Canal opened on the eve of the First World War, thereby providing BC's sawmillers with relatively cheap access to markets in northeastern North America and beyond. The province never looked back, and it quickly became Canada's dominant lumber producer – a title it still holds.[1]

BC's politicians were acutely aware that their longevity in office and the province's prosperity rested on a wooden framework, and so they devoted considerable attention and resources to supporting the industry. The provincial government controlled most of the woodlands that the lumbermen harvested, and it generated significant revenues from charging them a number of fees for cutting this "Crown" timber. By the early 1910s, in fact, BC's rosy fiscal situation was largely attributable to this source of income. In an effort to safeguard this precious industry, the province followed the lead of other jurisdictions in Canada by establishing in 1912 the British Columbia Forest Branch (BCFB) to administer and manage its woodlands. The BC government appointed soon-to-be legendary local lumberman Harold R. MacMillan, or "HR" as he became known, as the Branch's inaugural Chief Forester. Shortly thereafter, provincial authorities arranged for Canada's Commission of Conservation to conduct the first ever survey of BC's forests. It produced a report that concluded that British Columbia controlled roughly as much timber as the rest of Canada combined, an observation that only underlined the link between BC's trees and the health of its economy.[2]

But lest the province's forestry officials feel smug about their timber resources, they were being constantly reminded that their arboreal cornucopia came with many critters whose survival depended upon devouring BC's trees. No sooner had the BC government established the BCFB than the latter's officers began forwarding to the dominion's entomologists their concerns about the damage a variety of forest pests was wreaking on the province's forests. The issue had become so significant that BC's Minister of Lands, William R. Ross, wrote Gordon Hewitt, the Dominion Entomologist, in early 1913 to plead for the latter's assistance in tackling this vexatious problem. Hewitt responded enthusiastically to the inquiry. To tackle the issue he offered to dispatch a "specially trained officer," namely the recently hired Malcolm Swaine, who was in charge of forest insect investigations for the dominion government. Within short order, Swaine had communicated with MacMillan to discuss the logistics involved in conducting a forest insect survey in BC that season.[3]

MacMillan understandably saw this as a matter of the highest priority, and he expressed a remarkable appreciation for the forces that he felt were contributing to the problem and why it would only worsen as time passed; his views also reflected the contemporary understanding that the non-human world existed in a state of equilibrium. In writing to Swaine in April 1913, MacMillan did not mince words. He boldly insisted that British Columbia's bug issue was "the most important forest problem in Canada." To buttress his case he pointed out that BC was blessed with timber stands that were often pure, and the volume of wood per acre and stumpage values were relatively very high. "Admittedly," MacMillan surmised, "one half of the forest wealth of Canada is in this Province." Moreover, he predicted that the "reports of serious outbreaks of destructive insects" that he was receiving from his district foresters were probably the thin edge of the wedge. "It is quite likely," he astutely prophesied to Swaine, "that as logging increases in the future the balance of nature will be further disturbed, and insect attacks will be much more important than they have been in the past."[4]

MacMillan spiritedly facilitated Swaine's work by conducting the first insect study in British Columbia during the summer of 1913, a project he saw as presenting a unique opportunity for all concerned. MacMillan ordered his field officers to do whatever they could to support Swaine's work and cover part of its cost, and he asked the province's largest forest companies to do likewise.[5] MacMillan also publicized Swaine's study in North America's leading timber industry publications, including *Pulp and Paper Magazine of Canada*. Most importantly, MacMillan believed that the location of so much of BC's best timber along the province's coastline offered a singular opportunity in Canada to control insect infestations in commercial woodlands. As he put it, "Over the greater part of Canada, timber is so inaccessible that whenever it's [sic] destruction is threatened through insects, it is impossible to carry on logging operations in order to prevent the spread of the insects, or to salvage the timber. The conditions on the Coast are different," he emphasized. "The system of water transportation is so perfect, that large bodies of timber are equally accessible, and should an insect outbreak occur, it will be possible, in most cases, to have the timber removed without loss of time. Further, logging is carried on throughout the year, and infested or threatened timber may be removed at such time as is designated by an expert entomologist, whereas in other parts of Canada, logging is dependent on snow and weather conditions."[6]

Swaine produced a landmark report from the data he gathered during his field work in 1913 in BC. Significantly, it recommended that the timber companies take a number of simple control measures to tackle the beetle outbreaks they were confronting. These included harvesting or burning infested trees, floating the merchantable logs that they cut to drown the beetles, and burning the bark, slash, and slabs sawn from infested logs. To support his recommendations, Swaine cited evidence from the United States of instances in which the American entomologists had controlled beetle infestations by employing these different approaches.[7]

While Swaine's focus had been BC's commercial forests and the insects that were afflicting them, he had also been asked to investigate several pests that were causing havoc amidst the trees in Vancouver's fabled Stanley Park (Map 3). Among the world's cathedrals of urban green spaces, it stood out – and still stands out – as the Taj Mahal. It covers a roughly 1,000 acre, irregularly shaped jut of land that was originally covered in typical Pacific Coast rain forest, composed of (often giant) western red cedar, Douglas fir, western hemlock, and Sitka spruce. Prior to First Contact it was the site of a large Coastal Salish village, and Indigenous peoples cleared some of the trees to make room for their community. With sustained colonial settlement beginning in the mid-1800s, the Euro-Canadians sought to exploit the peninsula that comprises present-day Stanley Park for its military and economic value. The colonial officials set it aside in 1859 as a government reserve, but they allowed it to be developed thereafter. It became the site for a short-lived sawmill in the 1860s and the area was selection-logged of its best timber. A small number of settlers also created homesteads in the area, but most of the forest remained standing. Vancouver was incorporated in 1886 amidst a growing continental urban park movement, and the city's boosters eyed the government reserve in Vancouver's northwest end as an important part of their grandiose plans for their municipality's future growth. The following year, just as Frederick A. Stanley was appointed Canada's Governor General, they convinced dominion officials to lease the area to the city as a green space. As a result, both the ultimate symbols for hockey supremacy and for stunning urban parks in Canada bear Stanley's name.[8]

From the outset, local worshippers of Stanley Park in BC demonstrated a paradoxical attitude towards it that was widespread across North America and has been described as "natural constructivism."[9] On the one hand, they praised it because it exemplified untouched wilderness. To them, its august woods represented a vestige of natural beauty that was sacred and worthy of the greatest protection. As an editorial in a local paper put it in the early 1910s, "if Stanley Park is left as it is, it will remain alone among city parks, the only forest park in America actually in the front-yard of a big city." It was so delightful, the piece asserted, "because it is a tract of real British Columbia forest, almost untouched." On the other hand, the irony was that Vancouverites delighted in touching it in myriad ways in order to "improve" its appearance and render it even more "natural." While they enacted measures to preserve the park from development and exploitation, they permitted these "improvement" activities to proceed, but usually in ways that were disguised. During the late 1800s and early 1900s, for example, Stanley Park saw – among other things – some of its timber harvested and slash burned, its fowl hunted, one of its lagoons filled to convert it into a beach, and parts of its forest floor developed for tennis courts, lawn bowling, and playgrounds. Moreover, the municipal government built a ring road around the peninsula to render it more accessible to visitors, and a system of hydrants and water mains was installed to protect the area from fire. Although some development projects, such as a tramway and public stadium, were not realized, more than enough of them were, and they left no doubt that human finger and footprints were all over the park, even if onlookers failed to notice them.[10]

Unfortunately for those who revered Stanley Park, forest pests threatened to blemish its appearance on the eve of the First World War. Although it had suffered through an insect outbreak – and an episode of fungus – a few years earlier, these problems returned with a vengeance three years later. By mid-1913, locals saw the dead and dying trees with their de-leafed branches and browning needles in the park as an affront to its beauty, one that locals could simply not countenance. Although officials soon confirmed that the pests were all native and just as much a part of Stanley Park's forest heritage as its stately trees, this fact mattered not to those concerned with preserving its aesthetic appeal.

The members of Vancouver's Board of Park Commissioners, commonly referred to as the Park Board and established in 1889, knew which political levers to pull to get action with regard to their woody plant and pest problems in Stanley Park. On the provincial level, they turned to one of their local Members of the Legislative Assembly, Dr. George A. McGuire, for help. He dutifully raised the matter with William R. Ross, BC's Minister of Lands who had created the BCFB in 1912.[11] He directed its inaugural Chief Forester, Harold R. MacMillan, to study the situation in the park. MacMillan assigned George H. Edgecombe, one of his men, to this task.[12] Simultaneously, the Park Board conveyed its concerns at the national level through Henry H. Stevens, a former Vancouver alderman and its only Member of

Parliament; Stevens's Conservative Party had won power in 1911. He raised the issue with the dominion's Department of Agriculture, and thereafter pushed for help on the file.[13] No sooner had Stevens begun lobbying the Department than the Dominion Botanist, Hans T. Gussow, and head forest entomologist, Malcolm Swaine, were dispatched to investigate the situation, and Reginald C. Treherne, another dominion entomologist, was sent to Stanley Park to establish a field station there. The press covered the activities of these men closely, and even speculated on the diagnosis and treatment of the problem.[14]

The official reports on the situation were ready by early 1914, and they presented remarkably similar plans for defending the appearance of Stanley Park's trees against insects and pathogens; all entailed significant human intervention in its woodlands. Edgecombe, Gussow, and Swaine identified how its hemlock, spruce, and cedar were under siege. The most noticeable problem involved the hemlock, as dozens of them were dead and dying and had been rendered "extremely unsightly" by caterpillars.[15] Likewise, the park's grand cedars were being attacked by fungi, principally because much of the forest was clogged by dense underbrush that created a dank, dark microclimate that was an ideal breeding ground for them and destructive insects. In addition, the park's spruce trees were facing an onslaught from sap-sucking gall and green aphis. Once these trees were weakened, they were vulnerable to attack from beetles and wood-boring insects. To address these issues, the forester, botanist, and entomologist agreed that officials would have to closely manage the park over the long term. They recommended clearing the tangle of underbrush, cutting and removing the dead and dying trees (and ultimately even all the healthy hemlock and spruce), and replanting the areas that these trees had occupied with Douglas fir, which was much more resistant to diseases and insects. In addition, they called for the park's staff to cut the dead tops from the imposing cedars in an effort to spruce them up.[16]

Although these scientists also made one novel recommendation that aimed to preserve the infested but still healthy stands of hemlock and spruce, they recognized that there was a practical limitation to how high they could set their sights in this regard. They realized the uproar that would be caused if they simply went ahead and chopped down these mammoth specimens, particularly because so many of them lined Stanley Park's driveways and paths and were thus highly conspicuous.[17] So instead the scientists devised a short-term, triage measure, namely dousing the affected trees in a toxic spray of lead arsenate. When Swaine laid out the plan in early 1914 to MacMillan, however, he highlighted its unavoidable weaknesses:

> You can imagine what sort of proposition it is to spray those Spruce and Hemlock trees in Stanley Park. I don't know how high they are, but I would guess roughly 180 feet for

the larger spruce, and even the large ones are affected … Now we can spray all foliage under 100 or possibly 125 feet and I feel confident that we can control the outbreak on foliage below that height. We can obtain pumps run by gasoline engines which will drive a spray 100 feet from the ground … We could I feel sure save the trees within 1500 ft on each side of the roadway which would admit of a two horse truck. I shall be writing to the Park Commissioners again next week and shall give them my final recommendation.[18]

Swaine closed his report to MacMillan by highlighting how this effort would be unprecedented. "I have made all enquiries as to what work along that line has been done in the past, and can learn of no control work on such a scale," Swaine wrote. He thus proudly declared that "such an endeavour will be ploughing new ground." The alternative, he suggested, was to risk having practically every spruce and hemlock in Stanley Park fall victim to the bugs.[19]

There was significant support from a variety of interests for implementing this plan of action, with a different animus motivating each advocate. The media, for example, portrayed the bugs in the park as representing a devastating menace that could potentially ravage it. Naturally, dramatic measures were thus warranted. For his part, MacMillan conceived of the effort to undertake "insect control" in Stanley Park from a much broader perspective. He saw it as a microcosm of the general insect issues he faced in the province's commercial forests. As a result, he was grateful to have the advice of the dominion entomologists on matters of this sort, and thereafter offered them an even warmer welcome in an effort to deal successfully with the array of pests that sought to devour BC's valuable timber. He also urged Vancouver's Park Board to follow to a tee the advice that Swaine and his colleagues proffered to address the pest problems in Stanley Park.[20]

The upshot was a remarkable degree of cooperation among different levels and branches of government to realize the spraying project in the short term and the management of Stanley Park's woods in the long term. To monitor the forest pests there, Swaine had stationed one of his colleagues, Robert N. Chrystal, in Stanley Park, but he discontinued this effort after a few years when Chrystal fell ill.[21] Then, after determining that the dominion's Department of Militia and Defence would pay for the spraying operation in Stanley Park, Swaine's tiny team made history.[22] Chrystal's colleague, Reginald Treherne, oversaw the project, which occurred in March 1914, and it represented the first time in Canada – and probably North America – that such large forest trees were doused in deadly toxins in an effort to combat an insect infestation. In total, Treherne's team sprayed twenty-three trees using a machine that was shipped from Vancouver Island, where it

Triumphant

5.1 Cartoon of forest pest devouring trees in Stanley Park, 1914. Source: *Vancouver Sun*,
24 March 1914, 1.

was typically used to treat fruit orchards.[23] Although Swaine emphasized that the work was purely experimental, it created the impression that it would eradicate the pests from the poison-soaked trees.[24] Thereafter, crews began cutting and burning infested trees and planting Douglas fir seedlings in these disturbed areas. In addition, they cleared underbrush, which allowed them to reopen old trails and cut new ones.[25] Although some voices expressed fears that this work would be carried too far and that it was being done more for cosmetic purposes than as a means of insect control, it ultimately proceeded.[26]

The scientists and local citizens who were involved in and supported this work repeatedly stressed that their primary concern was maintaining Stanley Park's appearance; if its trees were not at their verdant best, they would lose all their value. For example, a concerned Vancouverite delivered an emotional appeal to the city's Park Board in the midst of this image crisis. He had warned that, if nothing were done, the park's trees would be destroyed and "they will become eyesores instead of ornaments, and the park will become a place of little or no interest or utility for many generations."[27] William Rawlings, Stanley Park's Superintendent, continually echoed these thoughts in justifying the work in Vancouver's most celebrated green space. For example, when he sought advice from BC's Inspector of Fruit Pests in early 1915 regarding how to combat the insects that were menacing the park's trees, he stressed that his goal was to devise the best possible strategy for securing "the preservation of the natural beauty, which is the charm of Stanley Park, and which after all is the object of us all."[28] Likewise, in his annual report for that year, Rawlings underscored that, "in dealing with this park of forest, as many people still prefer to term it, many features and viewpoints present themselves. Chief among them is the preservation of its natural beauty before all other considerations. Care has to be exercised in planning any new work so that it shall conform to the order of things as they exist, and that it shall preserve characteristic natural features."[29]

The hullabaloo surrounding the insect attacks in Stanley Park died down almost as quickly as it had sprung up, leaving the dominion forest entomologists in British Columbia to return to their true priority in the province. Specifically, Swaine and his colleagues focused on learning about the dendroctonus beetles that were devastating large swaths of BC's pine timber. A few more seasons of study – and an experimental control project with yellow pine near Princeton – convinced Swaine that he had solved the conundrum that was beetle control in this type of woodland. He authored two studies during the late 1910s that outlined his plan for realizing this aim, namely a robust program of burning infested trees and the stumps and slash created by logging operations. Money was too tight during the First World War for the provincial government to fund pilot projects to thoroughly test Swaine's theories, but the cessation of hostilities in 1918 brought him good news. Martin A. Grainger, who had replaced Harold R. MacMillan as BC's Chief Forester, agreed to allocate over $5,000 for an experimental, large-scale campaign to combat a beetle outbreak near Vernon in the wake of the overseas conflict.[30]

While Swaine was undoubtedly grateful for this news, he scored an even greater victory at this time. Ralph Hopping was a veteran entomologist with the United States Forest Service (USFS). Born in New York City in 1868, he had begun his university training there before he left for California because of ill health. He and his brother landed in the Kaweah Co-operative Colony in the Sierra Nevada Mountains; it disbanded when it was absorbed into the newly created Sequoia National Park. Hopping then established the park's first tourist operation, a business he ran until 1905 when his partner died and the USFS hired him. It had done so because he was an expert forest entomologist by this time, having studied insects since he was a young boy. As the USFS's forest entomologist in California during the early 1900s, he corresponded with colleagues literally across the world, including Malcolm Swaine in Canada, and he became renowned for his knowledge of bark beetles. Not only were they Swaine's primary focus as well, but they also represented the pre-eminent entomological threat to BC's enormous stands of valuable pine timber.

In Swaine's correspondence with Hopping, the former learned that Hopping was chafing under the practically impossible political situation that was plaguing the forest entomology bureaucracy in the US at this time, and Swaine seized upon his chance to add a true all-star to his staff. He offered Hopping the opportunity to oversee the dominion government's forest entomology work in British Columbia, and Hopping eagerly accepted the position and the offer to be Swaine's assistant. Swaine announced Hopping's appointment in November 1919, and a few months later Swaine correctly prophesised that, "now that we have Mr. Hopping with us, we can attack other important problems in the BC forests."[31] Hopping, whom Richard Rajala describes as having been "one of the foremost forest entomologists in North America,"[32] would be responsible for forest entomology in BC and western Alberta until his retirement twenty years later, at the age of seventy-one.[33]

Hopping led a dynamic and remarkably effective attack against forest pests in general and bark beetles in particular during the 1920s. The decade witnessed an extraordinary growth in BC's forest industry. Its productive capacity grew so much that American producers feared the competition from their northern neighbour, and thus ignited an incipient softwood lumber dispute by lobbying the US government to impose punishing duties on the Canadian product.[34] In terms of forest entomology, the more wood that BC's lumber industry harvested in the province, the more Hopping and his colleagues believed that it was upsetting the forest's "natural" balance and creating prime conditions for insect infestations. Hopping proved adept at pushing the BC government to implement new measures to tackle the province's most pressing forest pest problems. These included compelling industry to undertake control measures, such as harvesting infested trees before healthy ones and tackling the bark beetle problem; one particular initiative to control beetles in BC's interior proved to be an outstanding success.[35]

In addition, Hopping compelled the province's field foresters and rangers to report systematically on insect activities in their areas. The upshot was a remarkable level of cooperation between the dominion entomologists and the BC Forest Branch in dealing with insect issues in the province's commercial woodlands, as well as a slew of publications under Hopping's name. Fortunately for Hopping, he learned that provincial politicians had their own selfish reasons for being concerned about entomological issues. Wood borers had begun chewing their way through the floor of the BC Legislature by the mid-1920s, leaving the pests literally lurking beneath the feet of the province's elected officials, who reached out to Hopping for advice about how to defend against the siege.[36]

Although the entomologists and foresters remained focused on insect issues affecting BC's commercial forests throughout most of the 1920s, near the end of the decade their attention shifted to a far more glamorous entomological problem. It involved an American lumber company, a posh waterfront resort near Vancouver, a horde of tiny caterpillars, and various levels and branches of government. The team that confronted this pest problem would be led by an entomologist with a familiar last name, George R. Hopping, Ralph Hopping's son. As we have seen, Ralph came to Canada after the First World War to take charge of the dominion's forest insect work in British Columbia. Following in his father's footsteps, George graduated from Oregon State with a forestry degree in 1925 and was immediately hired by Canada's Division of Entomology to assist his father in dealing with BC's most exigent insect issues.[37]

One of the main players in this episode was the timber company Brittingham & Young (B&Y). It was based in Madison, Wisconsin, and overseen by President Thomas E. Brittingham and Secretary-Treasurer and General Manager Edward J. Young. Like many lumber firms in the American Midwest, most notably the iconic Weyerhaeuser Timber Company, its roots were anchored in the softwood forests of the Great Lake states. As the local supply of mature, commercial conifers receded, these enterprises looked farther afield for new timber stands. Whereas companies such as Weyerhaeuser opted to establish a major presence solely south of the forty-ninth parallel, B&Y chose to spread its reach north of the border as well. It incorporated an eponymous firm in British Columbia in 1912 with an office in Vancouver.[38] By this time it had already acquired a string of timber licences on the Indian River watershed, which flowed into the Indian – or North – Arm of Burrard Inlet in the eastern part of North Vancouver just across from the City of Vancouver (Map 3).[39] This location was unique in so far as it represented the divide between territory that was owned by the Canadian and BC governments. Land to the east of the Indian River was within the dominion's Railway Belt, whereas that to the west was provincial Crown land. As a result, B&Y obtained timber licences from both Ottawa and Victoria, and dealt with both governments to obtain permission to develop its holdings.[40] Within short order, B&Y had also acquired very large blocks of timber lands along BC's coastline, and was operating major sawmills near Prince Rupert and on Vancouver Island.[41]

Young was the real engine behind B&Y, and he appreciated trees for more than merely their value as timber. In the mid-1910s he purchased the Wigwam Inn, located in the estuary of the Indian River and near his local timber holdings. This luxurious, water-access resort had been constructed a few years earlier by an enterprising and quixotic local businessman, Gustav Constantin Alvo von Alvensleben. "Alvo," as he became known, was born in Germany and had immigrated to British Columbia in early 1904. Based in Vancouver, he created a vast network of businesses over the next decade, an endeavour that was apparently assisted by word that his family enjoyed tight ties to Kaiser Wilhelm. His financial empire collapsed at the beginning of the First World War, however, and the Wigwam Inn and his real estate holdings were taken over by the Custodian of Enemy Property, from which Young purchased the inn. After investing in the premises to increase its appeal, Young partnered with the newly formed Harbour Navigation Company, which he contracted to transport tourists from the wharf in downtown Vancouver to and from the dock at the Wigwam Inn. By the mid-1920s, the property was a well-known leisure attraction for the well-heeled in the Lower Mainland and beyond.[42]

To be sure, the Wigwam Inn was nestled in a splendid setting that was only a few hours away from downtown Vancouver's filth and grime. The Indian Arm was a narrow gorge that poked north from Burrard Inlet, and the inn was located near its head. The Coast Mountains gave the Arm's banks steep sides that were a sight to behold, covered for the most part in what seemed like an eternally green coating of diverse flora. The Arm was fed on a perpetual basis by the region's frequent mists and rains, and by the meltwater that cascaded down from the snow in the crags' upper reaches. Giant, ancient western red cedars and towering Douglas fir and Sitka spruce reached up to the sky, and an array of ferns and mosses provided a palette of green plants close to the ground. Logically, the inn's proprietors capitalized on the area's breathtaking environment in advertising it to the world, and the press repeatedly described it as "one of nature's loveliest gems."[43] A publicity piece in a Vancouver newspaper in the mid-1920s described the inn's environs as being composed of "blue mountains, tall evergreens, tinkling waterfalls, still seas – it has all the surroundings that make a summer resort ideal."[44]

It was completely understandable, then, that Young grew deeply concerned when the Wigwam Inn's star attraction – the local scenery – came under attack on

5.2 The Wigwam Inn, ca. 1925. Source: CVA 260–1194.104, photographer James Crookall.

the eve of the Great Depression. The dominion's forest entomologists had identified in July 1928 an acute infestation of western hemlock looper stretching far up the Indian River watershed, and the situation appeared dire. The insect had been indiscriminate in its gorging, feasting primarily on western hemlock but also the other species – both conifers and deciduous – in the inn's environs. "Practically all trees in the area are infested," *The Canadian Insect Pest Review* reported at the time, with "many hemlocks being completely defoliated." The situation remained critical for the rest of that year and all indications pointed to 1929 bringing even grimmer news.[45]

By that time, George Hopping had already seized the reins of the situation and would become the dominion government's de facto leader of the aerial campaign against forest insects in BC over the next few years. Just after the looper outbreak around the Indian River had been detected, George had begun tapping his – and his father's – entomological colleagues in the United States for information they had about using aircraft to drop poisons on pests that attacked trees. Although no one had even suggested at this point that this type of operation be undertaken in BC, George was laying the groundwork for the possibility of doing so, principally by benefitting from the lessons learned by those who had been involved in previous projects of this sort. He thus queried experts, such as Willard J. Chamberlin at Oregon State Agricultural College (now Oregon State University), for details about the best plane, hopper, and agitator to use when dusting forests.[46]

All the individuals who were concerned about the infestation in the vicinity of the Indian River were motivated to take action by the same impulse: the belief that the pests were ruining the appearance of the trees around the Wigwam Inn and that forceful action was needed to contain the damage. After an examination of the area in the spring of 1929 confirmed that the looper was continuing to ravage the local forests, George Hopping asked his dad for advice about how to proceed, specifically where to buy enough lead arsenate for such a project. George was adamant in this correspondence with his father that the infestation "would spoil the value of the Inn if there were many dead trees in close proximity" to it, but that a dusting operation would probably "save the conifers." He delivered the same message to the owner of the inn, Edward Young at B&Y, when he recommended in late May that B&Y participate in an aerial dusting campaign against the looper. This was needed, he emphasized, "if you wish to preserve the fine appearance of the ground surrounding Wigwam Inn." Otherwise, George added, the bugs would "cause most of the conifers in back of the Inn to become brown similar to those up the river." Young responded to Hopping's letter by explaining that he, too, was interested in taking action "to save the appearance of the property."[47]

Hopping soon learned, however, that he and Young held very different perspectives on whether an aerial dusting project, which must have sounded outrageously extravagant at the time, was the appropriate means to achieving their ends. Hopping was aware that Swaine and his colleagues were employing airplanes to combat insect infestations in eastern Canada. He thus saw the looper outbreak around the Wigwam Inn as an opportunity to demonstrate to BC's foresters, timber owners, and government officials that this approach could produce equally impressive results when battling his province's forest pests. But for Young, the principal behind B&Y who had only heard second-hand reports about the damage the looper was causing around his beloved wilderness retreat, his focus remained firmly fixed on the bottom line. After Hopping recommended that the insect outbreak be treated by aerial dusting, Young shot back with a litany of questions, all of which centred on the project's cost and who was going to foot the bill. In particular, he wondered if such efforts usually entailed "some sort of government co-operation."[48]

Hopping responded to Young's multiple queries in a way that betrayed the former's determination to see the dusting project around the inn go forth that year; in answering them, he made it clear that he had already mapped out the operation's scope. He explained that Young could ask Western Canada Airways, which had entered the aerial dusting business, how much it would cost to treat roughly 100 acres around Young's resort using approximately 2,000 pounds of calcium arsenate dust. He also directed Young to ask the BCFB if it would share the cost of the project. And if Young felt that the aerial dusting work was going to be too expensive, Hopping offered a rather daunting alternative: "Several high climbers with hand dusters could probably save most of [the] timber immediately surrounding Inn." Significantly, Hopping underscored that the clock was ticking for Young to make a decision regarding the project. "You have only forty days," Hopping declared emphatically to him. The dusting would have to occur before mid-July in order to be effective.[49]

At this point, Hopping focused his energy on marketing the project in an effort to foster the requisite political support for carrying it out. He wrote Robert C. St. Clair, the BCFB's District Forester for Vancouver, on 10 June to explain the situation. Hopping pointed out that Young had asked if the BCFB would be willing to pay a portion of the cost of the proposed aerial dusting operation around the Wigwam Inn. Although the damage the looper had caused was largely on the dominion government's side of the Indian River, Hopping slyly suggested that the province's forestry officials should be equally concerned about this pest's activities and willing to invest in combatting them. As he explained to St. Clair, "the hemlock looper is a very serious pest and could easily spread into Provincial timber. It would not be a bad idea to make this a test plot to study costs and feasibility of such a project on a larger scale."[50]

Hopping was now based at the Wigwam Inn and was doing all he could to push the venture forward as efficiently as possible, and so he predictably lost patience with Young's temporizing and his apparent unwillingness to heed the entomologist's advice. In making his case, Hopping also stressed that preserving the beauty at the Wigwam Inn was essentially a matter of dollars and cents. Young had indicated in early June that his lawyer, Joseph R. Grant, would soon be free to discuss the matter with Hopping. But Hopping had not heard anything from either Grant or Young by the last week of June, prompting Hopping to tear a strip off the two of them. "Apparently Mr. Grant has not taken the looper infestation in this place seriously enough," Hopping's excoriation of Edward Young began. After Hopping indicated that he had received a quote – under $1,000 – from a flying contractor that had agreed to carry out the aerial dusting project at the Wigwam Inn, he laid out his warning to Young: "If the Inn is worth anything at all $2000 would not be too much to spend to get rid of this infestation. Please do not delude yourselves that the trees up the valley will live. A large percentage of them are gone right now and there is a good chance that a considerable acreage of timber in back of the Inn will do likewise if immediate and effective steps are not taken to prevent it." This censure elicited a prompt letter from Grant indicating that he would be at the inn within days, but this news – and a bit more – only further infuriated Hopping. He unleashed another attack on Grant, and arguably for good reason. Hopping had heard that Grant was hoping to tackle the looper using conventional means, which again caused Hopping to deride Grant for his smugness and repeat his accusation that Grant was not taking the infestation seriously. "I understand that you contemplate bringing a spraying machine up here," Hopping wrote from the inn. "Please be advised that this machine will only be good for spraying the smaller trees nearer the Inn. The tops of the larger trees cannot be reached." Hopping tersely warned Grant that "a killing of the hemlocks over a large area behind the Inn will not make a pleasing background. The cedar also may be killed." Hopping closed by repeating that airplane dusting was the only possible way to treat "practically the entire area," and patronizingly lecturing Grant that "you cannot control an infestation such as this with toys."[51]

Simultaneously, Hopping was fighting the same uphill – and trying – battle with the BCFB. He felt it was disinclined to recognize entomological issues in the province's woodlands as bona fide forestry problems. As a result, while he was lobbying the BCFB to support the dusting project at the Wigwam Inn, he ingeniously seized the chance to leverage the whole situation into a teachable moment. Writing to Robert St. Clair, Vancouver's District Forester, in late June 1929, Hopping provided details about the spread of the looper infestation in the vicinity of the Indian River and urged the BCFB to see this matter as an exacting one that demanded immediate attention and resources. To buttress his case, he compared the havoc the insect was wreaking to the damage caused by another well-known forest hazard, one that governments across the country spared no expense in fighting. "The area up Indian River from the Inn is very much the same in effect as though a fire had gone through the area," Hopping pointed out. "Had this been a fire I have not the slightest doubt that the Forest Service would have spent about $4000 to control it if it had the same dimensions. It seems illogical to do everything in the one case and absolutely nothing in the other." He added that "airplane dusting is rather beyond the experimental stage" and that this technology had already proven its effectiveness in the war against forest insects. "You are almost sure to bump into these infestations in future," he warned St. Clair, "and now is the time to consider what you are going to do, if anything."[52]

As convincing as Hopping's appeal may have sounded, the BCFB remained sceptical as to whether it was worth investing in an aerial campaign against the looper around the Wigwam Inn. Alan E. Parlow, the Assistant District Forester for Vancouver, explained to Hopping in early July that he needed more details about the infestation, including a map and a "comprehensive report on [its] history and present progress," and the estimated cost of treating it. Parlow also reminded Hopping that the Indian River formed the boundary between lands belonging to Canada and BC, and that the "bulk of the infestation is on dominion lands." Parlow added that the matter boiled down to a cost-benefit analysis. "From the cursory examination I was able to make," he told Hopping bluntly, "if you add the damage done by this year's heavy infestation to the amount already killed there is doubt whether or not there is enough salva[gea]ble material left to justify control unless it can be shown that the infestation is apt to spread to adjacent valuable stands, and I am not aware of such stands in the neighbourhood." Parlow closed by underlining that the BCFB was not about to act hastily on this file. Only after receiving all the necessary data, he stressed, would he be in a position to recommend to his superiors in Victoria "this fall" whether they should support this operation. In other words, the idea of dusting at the Wigwam Inn that season was, to his mind, a non-starter.[53]

Parlow's dilatory attitude towards this issue greatly perturbed Hopping, just as Grant and Young's had earlier, and he felt compelled to deliver a stern lecture to the BCFB in an effort to enlighten it about entomological matters. Hopping certainly understood the need to provide accurate data and maps regarding the infestation in the Indian River valley, and he gladly did so, along with information about the extensive aerial dusting work that had gone on in eastern Canada. And he recognized that the previous year's infestation had been mostly on dominion land, but he cannily pointed out that that was "begging the question a little. The shoe may be on the other foot this time." Hopping added that Canadian officials were discussing the dusting project and that the cost "would naturally be apportioned in direct ratio to the affected areas under the jurisdiction of each branch in conjunction with private owners." But that was beside the point, he stressed. "The present infestation is not worrying me so much as the attitude of the Provincial Forest Branch toward such things," Hopping's admonishment to Parlow read. "The idea seems to be that such infestations are not straight out and out forest protection problems, the same as fire. If one of your rangers failed to cope with a fire as near as this is to tide water and to Vancouver," he surmised, "and gave as his excuse that the timber was very poor quality, I think that you would at least give him a severe reprimand. If such an attitude toward severe insect defoliation continues there is sure to be severe criticism from timber interests." Hopping's peroration included some cutting remarks. "You have equipment in constant readiness for fires," he highlighted, "why should you not have equipment, or some arrangement with an airways company which could supply equipment, to cope with the depredations of defoliators?" Hopping explained that he was only waiting for a bit more information from back east about the results of the aerial dusting work there before recommending to the BCFB that it obtain similar equipment for use in waging war on forest insects in British Columbia. His final line spoke to the depths of his frustration with the present impasse. "Then the Forest Branch cannot ask us that question which it has worn to a frazzle," he chided Parlow, "'Well, what can we do about it?'"[54]

Before the BCFB even had the chance to pose this rhetorical question, political forces left it no choice but to become a central player in this affair. By the time Hopping was done lambasting Parlow, the owner of the Wigwam Inn, Edward Young, had shrewdly played a most powerful card. After being castigated by Hopping for not taking the inn's insect problem seriously enough, he had immediately dispatched his lawyer in Vancouver, Joseph Grant, to the site to tour it with Hopping and to witness first-hand the havoc the looper infestation was wreaking there. This was all it had taken to spur Young to take definitive action post-haste. He ordered Grant to travel to Victoria to see Peter Z. Caverhill, BC's Chief Forester, about the situation and demand that the province come on board in addressing it. As much as Young was the Wigwam Inn's proprietor, he was more importantly one of the province's leading lumbermen. Because British Columbia's economic backbone was made of wood, when the industry's leaders spoke, the province's politicians listened. Not surprisingly then, no sooner had Young's local representative visited Caverhill than the BCFB became a fervent supporter of the plan to conduct an aerial dusting operation around the Wigwam Inn. As George R. Hopping delighted in reporting to Malcolm Swaine back in Ottawa after learning of Young's intervention in the issue, "we are aided in our presentation of the case by several large timber owners who are at present going after the powers that be hammer and tong."[55]

Hopping immediately recognized the crushing political weight that had been brought to bear on the BC government (and apparently officials responsible for administering dominion lands in BC), and he was optimistic that it signalled the dawning of a new era in terms of how the province approached insect issues in its forests. Undeniably, officials in British Columbia had supported the dominion entomologists' efforts to tackle beetle infestations in the province's commercial pine woodlands, but the former had been most resistant to accepting that other forest pests represented a threat to the health of BC's timber. Resistant that is, until now. "We have now arrived at the turning point in the attitude of the BC Forest Branch toward such things as spruce bud worm and hemlock looper infestations," Hopping gleefully told Malcolm Swaine in early July 1929. Hopping now believed, for the first time, that the BCFB would conceive of these matters as being strictly ones of "forest protection" that "must be met instead of being ignored as in the past." He based his observations on the letters he now had from both the BCFB and the Dominion Forest Branch (which oversaw Ottawa's timber holdings in the province) "expressing willingness to do what is feasible to suppress these outbreaks." For good reason, then, Hopping felt that the time for conducting an aerial dusting operation in BC "was never more opportune."[56]

And opportune it was. Hopping frantically tried to make all the arrangements for the project at the Wigwam Inn to proceed that season, and in the process he enjoyed the BCFB's unquestioning support. He had already begun taking care of the project's practical details a few months earlier, though his frustrating search for 2,000 pounds of calcium arsenate dust had resulted in him obtaining only about half this total.[57] But by early July he had also filed the detailed report and maps for which the BCFB's officials had asked. At this time he also wrote Swaine in eastern Canada and urged him to send all the information that was available on the aerial dusting operations that were being conducted there. Most importantly, Hopping made one major request of Swaine, namely that the latter send the trimotor Ford aircraft, which was proving so successful in dusting the hemlock looper near Baie Comeau, to BC once it had completed its mission in Quebec so that it could be used to conduct the effort at the Wigwam Inn. Once again, Swaine bypassed his Deputy Minister in asking James Gordon, the Air Board's director, to fulfil Hopping's wish. This time, Gordon shot Swaine down, telling Swaine that his request "is not concurred in."[58] Hopping was thus left to find a plane locally that could carry out the project, and fortunately he had already lined one up through Western Canada Airways.[59]

These circumstances, and a few more that were about to develop, left Hopping as a virtual one-man band in overseeing the inaugural aerial dusting operation against forest insects in western North America. While Swaine and his entomological colleagues back east certainly conveyed as much advice as they could, Hopping was forced to reinvent the wheel in carrying out this work because so much of it was new to him and, frankly, new to the Canadian aerial dusting campaign. For starters, Western Canada Airways had agreed to fly the sorties at the Wigwam Inn, but it would do so using a plane – the Boeing B1E Flying Boat – that was at first sight completely unsuited to performing this mission. Just as its name suggests, this

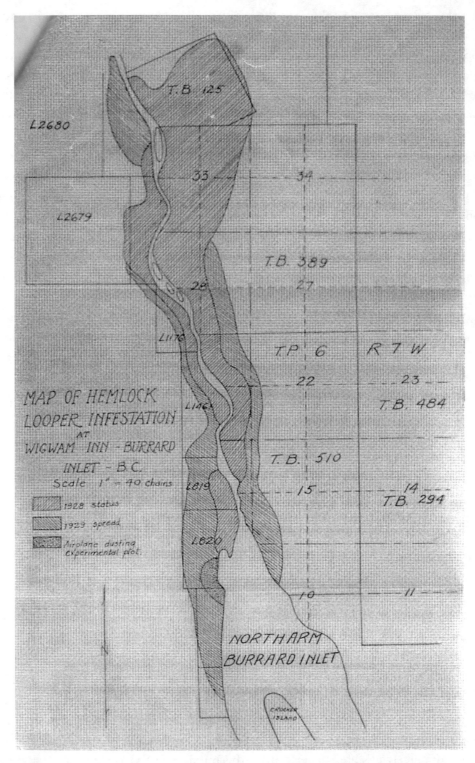

5.3 Map of hemlock looper infestation and area dusted at Wigwam Inn. Source: G.R. Hopping, "The Western Hemlock Looper" (MA thesis, Iowa State College, 1931), 71. Courtesy of University Library, Iowa State University of Science and Technology.

5.4 Boeing B1E Flying Boat taxiing before dusting Wigwam Inn. Source: LAC, RG39, 341, 47278, enclosed with 13 June 1929, J.W. Swaine to J.R. Dickson (e002107477).

aircraft took off from and landed on its "belly," a characteristic that made sense during a period when there were few airports and for flying missions in regions that were proximal to water bodies. With its fuselage in the water, the aircraft was kept upright by a stabilizer at the end of each of its wings.

In contrast, all of the planes that Swaine's crew had used in the dusting operations in eastern Canada had landed on pontoons that allowed their fuselages to sit well out of the water. This was a crucial consideration for a dusting aircraft because it released its poisonous payload through a hopper in the floor of the fuselage. For the Boeing Flying Boat, this meant that its hopper would protrude down into the water each time it landed and took off. In fact, the chief aeronautical engineer for Western Canada Airways doubted that it would be possible to distribute "dust from a flying boat such as ours." Moreover, it also meant that the hoppers that Canadian scientist had built for the aircraft that had conducted the dusting work in eastern Canada could not be used in BC.[60] Hopping thus had to intensify his pioneering effort and configure a hopper that would work on the Boeing Flying Boat. The RCAF's draftsmen offered him as much counsel as they could, but again, they had no experience deal-ing with this particular engineering challenge. Ultimately, Hopping resolved the issue by contracting a metal workshop in Vancouver to construct both the hopper and the internal agitator to keep its dust moving smoothly through the device. Sig-nificantly, the BCFB agreed to pay the bills associated with these expenses.[61]

Compounding the challenges Hopping faced was the passage of time. He had warned the Wigwam Inn's owner a few weeks earlier that the dusting would have to occur before mid-July or else the hemlock looper would be too far advanced in its development to be affected by the poison dust. Hopping was thus feverishly trying to address the litany of practical challenges attendant upon carrying out this experimental enterprise. But by mid-July, the hopper was not ready. As a result, Hopping had to scale back the project from an all-out effort to control the looper to one that would merely test out the method on a smaller area. On 19 July, Hopping anxiously wrote Major Donald MacLaren, the head of Western Canada Airways, and advised him that the looper larvae were growing quickly and would soon reach their adult stage, by which time "it will be too late to dust." He thus pleaded with MacLaren to let him know as soon as possible when the Major would be ready to proceed with the project.[62]

But then, Mother Nature intervened to extend a helping hand to Hopping and his trailblazing undertaking. Although the summer in Vancouver typically brought drier weather, in 1929 it witnessed a prolonged period of rain in the Lower Main-land that sent the loopers' development in the Indian River watershed into abey-ance. Hopping could not contain his excitement in relaying the good news to the BCFB. His letter of 24 July to the District Forester's office in Vancouver explained that the recent bout of liquid sunshine had "acted to greatly slow up the feeding of the larvae on the Wigwam Inn area and that the great majority of them have not yet reached more than three fourths of their full size." This meant that there were still roughly a dozen days remaining for dusting that season.[63]

After a few more delays, Hopping achieved his aim of conducting the aerial dust-ing project around the Wigwam Inn, albeit a very limited one. While the hopper was ready to go by late July, the continued wet weather prevented the effort from commencing. In addition, Western Canada Airways' Flying Boat had to fulfil its other contractual obligations, including local fire patrol duty, before it could under-take its dusting mission for Hopping. But on 30 July it was freed up and able to do so. That evening, as well as the next morning, it spread roughly 1,200 pounds of calcium arsenate on approximately forty-five acres behind the Wigwam Inn (an average of twenty-six pounds per acre). Major MacLaren piloted the plane and lau-rels were heaped on him for his expert flying ability. Similar praise was directed at Hopping, who flew with him. Both turned the crank on the agitator to ensure the

5.5 Dust falling around Wigwam Inn. Source: LAC, RG39, 341, 47278, enclosed with 13 June 1929, J.W. Swaine to J.R. Dickson (e002107478).

5.6 Plane releasing dust near Wigwam Inn. Source: LAC, RG39, 341, 47278, enclosed with 13 June 1929, J.W. Swaine to J.R. Dickson (e002107476).

smooth flow of the toxic dust through the hopper, and together they opened and closed its door as the plane flew over the drop zone.[64]

George Hopping was absolutely ecstatic when he began wandering through the forest that had been treated behind the inn to view the results of the dusting. In fact, he was downright delirious when he realized that he had entered a virtual killing field, and he could not contain his excitement. Less than a week after the airplane had dumped its deadly payload around the Wigwam Inn, Hopping forwarded the preliminary data from the dusting to the BCFB's District Forester in Vancouver and added the obvious: that "a great many larvae have been killed."[65] A day later, Hopping wrote B&Y in Wisconsin to report that the dusting project had been completed and that he could "not hesitate to state that it is and still continues to be an

5.7 Plane dusting shoreline near Wigwam Inn. Source: LAC, RG39, 341, 47278, enclosed with 13 June 1929, J.W. Swaine to J.R. Dickson (e002107479).

unqualified success." He relayed the same wonderful news to Joseph Grant, B&Y's local solicitor, and added that he estimated that 85 to 90 per cent of the larvae had been destroyed.[66] Next he shared news of his success with his father Ralph Hopping, who was Canada's senior forest entomologist in BC. As George boasted to his dad, "the dusting was even more effective than I thought that it would be. The larvae died by the thousands." He added giddily, "incidentally, I have done considerable flying."[67] Finally, Hopping would have told the Dominion Forest Branch (DFB) about the project's success, but its own personnel beat him to the punch. Ed Walmsley, the DFB's superintendent for the Coast Forest District, elatedly told Hopping that one of the DFB's officials had just returned from inspecting the dusted area and described how "the work has been successful, in fact more so than [he] had anticipated owing to the somewhat crude equipment available for dusting."[68] Overall, Hopping would subsequently proclaim that "the forest trees around a summer resort at Wigwam Inn on Burrard Inlet were saved from destruction by this control operation."[69]

Anxious to share his extraordinary results with a broader audience, Hopping produced a preliminary report about the aerial dusting project at the Wigwam Inn by early August and it was telling for several reasons. First, it betrayed his steadfast belief that the project could have yielded even better results had B&Y been a more amenable partner. Hopping's report recounted how he had first contacted the firm in late May to ask for its assistance in tackling a very serious insect infestation at the inn, but that unfortunately the firm had neglected the matter until early July. This had left very little time to prepare for the project and thereby turned it into, as he put it, "an emergency measure, organized and completed in less than a month."[70] As a result, he argued that future efforts would be even more effective if they were carried out according to a proper schedule. Second, Hopping stressed that the project was truly a pioneering affair that had taught those involved a host of lessons, including the fact that the Boeing B1E Flying Boat was deficient for this sort of work. Hopping was adamant that in the future he use a bigger plane, one capable of carrying a minimum payload of 1,000 pounds of poison and "fitted especially for dusting," and that doing so would greatly lower the cost of the work.

In addition, Hopping pointed out that he lacked a scientific method for determining the percentage of mortality among the target insects, and in this regard he was not alone; no one had yet figured out a reliable means for collecting this data. One strategy had been to place a drop cloth underneath the treated trees and to count the dead caterpillars as they plummeted to the ground, but this had proven ineffective. Although the caterpillar larvae would fall to the earth as the chemical dust started taking effect (normally they would lower themselves on the "threads"

5.8 Mass of hemlock looper larvae after aerial dusting at Wigwam Inn. Source: G.R. Hopping, "The Western Hemlock Looper" (MA thesis, Iowa State College, 1931), 60. Courtesy of University Library, Iowa State University of Science and Technology.

they spun), they would just as quickly pick themselves up off the drop cloth and instinctively start climbing the trees again.[71] Ultimately, they would "accumulate in huge masses on stumps and roots" where they would die. Hopping thus suggested two alternative methods for calculating the mortality of the target bugs, both of which reflected how the innovators involved in this work were as much artists as they were scientists. The first involved comparing "the density of the droppings both before and after dusting." And he explained that this should be supplemented by a second method, which would compare "the volume of sound produced by the falling particles before and after dusting." He noted that "before dusting this area, the droppings sounded like a steady rain and now the sound of droppings has almost ceased." In his eyes, or more specifically to his ears, the conclusion was elementary: "this dusting has been a great success from the standpoint of the results."[72]

Lastly, Hopping stressed that it was axiomatic that the aerial dusting project had been worth it because, unlike a stand of commercial timber, nature's aesthetic value was incalculable. Predictably, all participants were keen to tally the project's cost, for this burden was divided equally among the owner of the Wigwam Inn, the DFB, and the BCFB. But in calculating the effort's total price tag – it came to just over $470 – there was no debate over whether it had been worthwhile. This was in sharp contrast to entomological treatments that occurred back east in commercial woodlands, whereby the decision to conduct a project was based upon a careful consideration of its cost versus the value of the trees that would be protected. In the instance of the Wigwam Inn and the dusting that had occurred over 1928–9 in Muskoka, Ontario, however, everyone tacitly agreed that such computations were simply unnecessary when it came to, as Hopping's report put it, "the purpose of saving the scenic appearance of a valuable summer resort."[73]

When Hopping reflected upon this operation, however, his views exemplified the inherent contradiction that lay at the heart of society's attitude towards nature. Even when he admitted that preserving the aesthetic value of the woods in this instance was really all about money, it was, he believed, an act upon which he could not place a price tag. In a later analysis of the project, he asserted that the "Wigwam Inn is a summer resort, valued at not less than $100,000, and it was extremely desirable to preserve the scenic beauty of the immediate Inn surroundings, irrespective of the timber values involved." In fact, he declared, there was no need to worry about the cost of the operation on a per acre basis "because the value of the timber, as such, was not involved."[74]

As much as the project created a strong buzz among the entomologists and foresters, it also caught the attention of the media and opportunistic chemical firms near and far. The *Vancouver Daily Province*, for instance, published a story immediately after the dusting project wrapped up and described the effort as having been "entirely successful."[75] Likewise, the editor of the leading industry news magazine, *West Coast Lumberman*, heard about the project at the Wigwam Inn even before it began and eagerly solicited information about the campaign from those conducting it.[76] J.H. Boyd, president of the Commercial Chemical Company in Tennessee, also wrote the BCFB after Hopping had concluded the dusting in an effort to solicit future business for the calcium arsenate it manufactured.[77]

But some of the personnel involved in the project at the Wigwam Inn were not caught up in the excitement of the moment, and instead they disparaged it for what they saw as its obvious shortcomings. Ed Walmsley, the DFB's superintendent in the Coast Forest District, had been an enthusiastic supporter of the project from the outset, and his zeal only grew after its stellar outcome became apparent.[78] He wrote his superiors – Charles MacFayden, who was the DFB's District Forest Inspector in BC, and Ernest H. Finlayson, the Director of Forestry back in Ottawa – with the good news. After describing the outcome as "excellent" and stating that he was "more than pleased with the result," Walmsley added the following presumptuous prediction in his letter to MacFayden: "While you were sceptical I have no doubt you will be gratified to learn of the large measure of success obtained."[79] But Mac-Fayden certainly was not gratified, and he pulled no punches in forwarding his critical views to Finlayson. MacFayden stressed to his boss that his job was to manage the dominion government's timber throughout British Columbia, including the valuable stands in the Indian River watershed that were both infested and in danger of becoming so. But the dusting project around the Wigwam Inn had done virtually nothing, he argued, to protect this wood, and instead had essentially been a self-serving project that the inn's owner had promoted to protect the arboreal vista for the inn's wealthy patrons.[80] As MacFayden sarcastically described the operation that had just taken place, the dusting had occurred in "the timber immediately surrounding the Wigwam Inn with a view to saving it for its aestetic [sic] value and in doing this the primary object would be to kill the caterpillars, even at a cost that would be prohibitive on a larger scale."[81] In other words, the DFB had wasted its money in agreeing to pay one-third of the project's cost.

MacFayden then provided Finlayson with a bit of background on the story, and in the process assailed those involved in the effort. He explained that he had opposed the project at the outset and had doubted the ability of its advocates to improvise the equipment needed to pull it off. "Rather than be the only wet blanket on the whole scheme," he admitted, "I consented to bear an equal share of the cost." And yes, the project was undeniably successful in terms of killing the looper around the Wigwam Inn, but as far as experimental work was concerned, Mac-Fayden was adamant that "very little has been gained." Dumping 1,200 pounds of calcium arsenate on forty-five acres killed "80 or 90% of the caterpillars," he underscored. "These facts, however, were previously known and the experiment, if it can be called such, gives no new information. For instance," he continued, "it is yet a matter of conjecture as to whether or not one third or three quarters of the arsenate would have done the same work, and it is still unknown what quantity would be required to kill 100% of the caterpillars. In other words," he declared in summary, "if this was to be properly carried out, it is necessary that the proper equipment be available and laid out in plots as done last summer by Doctor Swaine in the East."[82]

The officials in the vanguard of the forest entomology campaign in Canada had a forceful rebuttal to MacFayden's criticisms, however, for in their opinion, his carping had completely missed the point of the exercise. To be sure, Swaine recognized some of MacFayden's points in discussing the matter with D. Roy Cameron, a senior forester with the DFB. "It is true," Swaine openly conceded to Cameron, "that our two extensive experiments for the hemlock looper in the East this year were very successful and have given us practically all the information we need to guide us in commercial airplane dusting." But that was irrelevant, Swaine emphasized, because the merit of the project at the Wigwam Inn was not in its efficacy but in its publicity. As he put it, "I feel that Mr. Hopping's small operation in British Columbia was of great value in demonstrating to the people there that this work is actually feasible."[83] Similarly, Ed Walmsley with the BCFB zealously defended the project to Cameron, but did so in an even more maudlin manner:

> I can not agree with Inspector MacFayden's statement that from an experimental standpoint very little has been gained. In fact, to be honest with ourselves I think we all learned something of real intrinsic value, and I can not recall any extraordinary expenditure made by this Service during the past twenty years which could be more readily justified. As a matter of fact, I was so enthusiastic about the project that if inspector MacFayden had refused to approve of the financial commitments which I made, I would have paid the money myself.[84]

By the time of Walmsley's dramatic declaration, however, everyone concerned with the health of British Columbia's forests was grappling with much more important issues. Near the end of the summer of 1929, reports from foresters and entomologists alike had confirmed that BC's Lower Mainland was now host to a patchwork of major hemlock looper infestations (Map 3). Ignoring them was not an option, most especially because several of them were centred in areas that were considered too valuable to be laid waste by the pests' depredations. The story of how officials dealt with two of these infestations is the final topic to be addressed.[85]

CHAPTER SIX

"Carrying out this work, of a protective nature"

COMBATTING FOREST INSECTS IN SEYMOUR CANYON AND STANLEY PARK,
BRITISH COLUMBIA, 1929–1930

As Canada and much of the world headed towards grievous economic dislocation in late 1929, a remarkably diverse array of officials in British Columbia were grappling with their own unsettling situation in the province's forests: a host of insects was munching its way through stands of valuable timber in different parts of BC. Although these pests were setting off alarm bells in both the public and private sectors with the destruction they were causing, even broader attention was drawn to a particularly worrisome series of infestations. The culprit was the hemlock looper, and the outbreaks were peppered throughout the Lower Mainland. Ralph Hopping, the senior dominion entomologist in the province, identified the spectre in his annual report for 1929 as "threatening hemlock on the Vancouver watershed, Stanley Park and several Provincial areas of fine pulp timber." As a result, he predicted that "operations for control will probably be extended in 1930, and several areas will probably be dusted by airplane."[1]

Two sites in BC would be treated that season, and the reasons why coincided almost perfectly with those that had precipitated the projects in Muskoka and around the Wigwam Inn. Business and political leaders in Vancouver in particular and British Columbia in general conceived of both Seymour Canyon (and by association Vancouver's drinking water) and Stanley Park as being indispensable to the city's identity. In their eyes, integral to maintaining and enhancing the municipality's continental and international profiles was the need to protect the sanctity of these two cherished locations. To be sure, both sites exemplified "hybrid nature" and "organic machines," wherein humans had significantly altered these environments in an effort to improve them and yet were blind to the myriad ways in which they had done so.[2] Consequently, when a forest pest attacked some of the trees in these locales and they were seen as losing their beauty as a result, it was hardly a stretch for their boosters to embrace wholeheartedly aerial dusting with toxic chemicals as a means of restoring the woods to their former glory. In the process, the proponents of these operations argued that it was nearly impossible to place a value on the trees that were being protected because they were admired for their appearance and not their timber qualities. In addition, a remarkable esprit de corps guided the actions of the participants in dusting Stanley Park and Seymour Canyon in 1930. Officials from a number of branches of the dominion, provincial, and municipal governments collaborated to realize these projects. Finally, the effort around the Wigwam Inn the previous year had highlighted that BC was operating in a distinct universe of its own. The mountains and vast expanse of land that separated the province from the rest of Canada created colossal practical barriers to cooperation between East and West. This left those on the country's Pacific Coast to fend largely for themselves when it came to figuring out how best to bomb their looper infestations with toxins.

In terms of aerial dusting in British Columbia, George R. Hopping had been and would remain its driving force in the province. He had almost single-handedly carried the project that had occurred at the Wigwam Inn in 1929, and he was so impressed by its results and driven by his desire to learn more about this technology that he was profoundly committed to conducting another operation, if not several of them, as soon as possible. Furthermore, he was acutely aware of the many challenges he had overcome to achieve his aim, and he was driven to line up the necessary political backing to realize his agenda in 1930.

A crucial element in this plan was drawing the attention of important elected officials and bureaucrats to the pressing need to combat the menace that the hemlock looper represented to their aspirations, and in this regard Hopping succeeded with aplomb. Although the dominion entomologists had been tracking a number of outbreaks in the Lower Mainland, Hopping knew that several infestations were located in sites that were considered, for very different reasons, near and dear to the hearts of Vancouverites. Presumably, he must have reasoned, a suggestion that the insect epidemics in these particular sites could be eradicated by aerial dusting would gain the backing not only of BC's forestry officials but also, and perhaps more importantly, of the city and province's most influential boosters.

And so Hopping set out to implement this plan. He strategically focused on a few of BC's most pressing forest entomology issues in a broad-ranging report he prepared in the fall of 1929. The most intense outbreak was still in the Indian River watershed where he had worked the previous season, but he and his field staff had located a handful of new "areas of infestation." One involved the hemlock looper in Stanley Park, and it had already defoliated some of the park's most stately trees. "It is very important," Hopping declared in this report, "that the Stanley Park timber be maintained in its present state." As a result, he began surveying this entire green space to get a better understanding of the problem, and promised to send William Rawlings, its superintendent, an estimate of the cost to control the outbreak. If the looper expanded its range of activities in the early spring, Hopping declared that "it will be a question of airplane dusting the large timber blocks, and using a ground sprayer along roads, paths and in the recreation areas."[3]

Hopping added that the second most severe outbreak of the hemlock looper that he had detected was in Seymour Canyon (or Creek), which flowed through North Vancouver and emptied into the Burrard Inlet just west of its North (or Indian) Arm (Map 3). He recognized that officials would consider this particular infestation in the most serious light because it was afflicting a watershed that formed a crucial part of Vancouver's water supply. Hopping described how the hemlock looper outbreak extended "along Seymour Creek from a quarter of a mile below the city reservoir, up the canyon as far as Loch Lomond, a distance of twelve miles." He promised to discuss the situation and possible control measures with officials from the Greater Vancouver Water Board.[4]

In relaying this news, Hopping had skillfully jabbed two ultrasensitive nerves in the collective consciousness of Lower Mainlanders. First there was the issue of forest pests in Stanley Park, which the preceding chapter has already demonstrated was a sacred place for many Vancouverites. Roughly one and a half square miles in size, it occupies a promontory of coastal rain forest that pushes out into Burrard Inlet to create what is known as the First Narrows. Initially the site of a Coastal Salish village, the area experienced limited industrial development after First Contact before being formally established as an urban park in 1887. Thereafter, its worshippers brought a strikingly paradoxical attitude to it. Although they exalted the park, especially its defining giant trees, for representing untouched wilderness, they simultaneously manipulated it in an effort to improve its aesthetics.[5]

Stanley Park was cherished as a marquee attraction in Vancouver, an urban forest oasis the scope of which was unmatched on the continent, if not the globe, and its allure arguably increased in step with urbanization and industrialization during the 1920s. "Stanley Park is not only one of the chief scenic attractions of Vancouver, but it is one of the most unique parks on the continent with a world-wide reputation among travelers," an article in the *Christian Science Monitor* explained to its readers in April 1926. "Its chief charm for both visitor and resident," the piece continued, now aptly capturing the enigmatic manner in which people conceived of the park, "is that the greater portion of the park is still primeval forest; and the utmost care is taken by the Vancouver Park Board to see that its countless natural beauties are not disturbed. One may step off a street bustling with the activities of a modern, growing western seaport and within five minutes be alone in the depths of a fine old forest, where the only sounds are the rustling of the leaves and the twitter of birds." It was the home "to no ordinary forest," the article repeated: "The home of big trees, the Pacific coast boasts few larger trees than are to be found on the peninsulate, all of which is occupied by Stanley Park, and which is so carefully guarded against fire or destruction by human agency. Here are to be seen Douglas fir and cedar rising 300 feet in air, and many of which are established by foresters to be 700 years and more in age."[6]

As alluded to in the foregoing article, Stanley Park's champions repeatedly felt justified in taking measures to protect and improve its appearance, and never felt that doing so detracted from its vestal quality. During the 1920s, for example, such measures included fighting lengthy legal battles to clear the park of its "squatters," killing its unwanted pests – such as earwigs – with poisoned baits, performing landscaping work to beautify it further, and stocking its lakes to improve their attraction to local fishers.[7] Naturally, when forces worked to desecrate its "primitive beauty and wealth of foliage and giant trees," as one article put it, corrective measures were needed. In describing the defoliation wrought by the hemlock looper in Stanley Park during the late 1920s, for example, the same piece decried how the resulting "dismal skeletal aspect of the trees was offensive aesthetically."[8]

Vancouver's water supply, or more specifically the environment from whence it came, held an equally cherished place in the hearts and minds of the city's leaders. The City of Vancouver and its surrounding municipalities were unique in North America, and perhaps in nearly every major urban centre on the planet, because of the purity of their drinking water. Practically all the world's cities were compelled to treat their potable water in various ways, an approach that almost always entailed countering its impurities with chemicals – most often chlorine – in order to render it safe for human consumption. This reality was most often a function of the urban centres having been established and developed on or near a water source that they had contaminated as they had expanded. Cities were thus were forced to attempt to undo the damage they had caused by using synthetic additives to render their water safe for drinking. Not so in the Lower Mainland. There, the water emanated from a few rivers that snaked their way through the Coast Mountains that formed a northern bookend on metropolitan Vancouver's growth and emptied into Burrard Inlet. One of these waterways was the Capilano River and the other was the Seymour River (it was referred to as both a creek and a river, so both terms will be used herein). Because both were areas of virtually untouched wilderness, they were considered to epitomize purity, and so all that was required before their contents could be piped into the homes of Vancouver's citizens was a simple filtering (Map 3).[9]

As a result, many of the Lower Mainland's boosters in the early part of the twentieth century felt that Divine Providence had preordained their urban centre to soon rank among the continent's greatest metropolises. The city's media repeatedly trumpeted that "Vancouver has many natural advantages as a city, and one of the most important of these is plenty of pure fresh mountain water and its proximity to its source."[10] Similarly, the province's largest daily declared triumphantly in the mid-1920s that, "with an adequate supply of the purest fresh water, Vancouver is second to none in this important factor to the prosperity of every community. Coming direct from the snows and glaciers of the mountains across the Burrard Inlet to more than 50,000 homes in Vancouver, it answers the turn of the tap, crystal-pure and sparkling."[11] Finally, Lower Mainlanders demonstrated that, even as early as the "Roaring Twenties," they delighted in showing up Toronto, in this case by comparing the quality of their water supply to that available to residents of the Queen City. A rather snide editorial in the *Vancouver Daily Province* underscored just that: "We needed water and we found it at our door while other cities have had to go hundreds of miles. We wanted pure water and we went to nature's own laboratory and had it distilled for us. Other cities have had to spend millions on filtration and chlorination plants and then have got not so much water as a chemical mixture. We supplied pipes, and Mother Nature's own gravitation engine did practically all the pumping that was necessary." In contrast, Toronto had to treat its water supply with a double process of filtration and chlorination. Furthermore, the city had recently decided to increase the concentration of chlorine it used in this treatment in order "to kill the taste of phenols" and also add "another drug to neutralize the taste of the additional chlorine." The editorial closed by rhetorically asking its readers, "Can Vancouver citizens imagine themselves drinking this 'harmless' combination of phenol, chlorine and some other drug, and calling it water? How would they like to trade their Capilano brand for it?"[12]

Vancouver's luminaries were acutely aware that they had been blessed with a water supply that made their city nonpareil on the continent, if not around the globe, and they thus did all they could to keep these watersheds immaculate. In the eyes of the city's boosters, these sites would only retain their sanctity if they remained pristine, specifically by showing virtually no traces of human activity. To realize this end, and to manage the water supply in the Lower Mainland, Vancouver's municipal leaders attempted to convince provincial politicians to transfer control over the watersheds from which they drew their drinking water to the civic officials. While it took some time, they eventually succeeded in this endeavour, and in their efforts to compel the provincial government to enact a series of laws that tightly governed human activity in these watersheds. The new statutes required those who wished to recreate or work in these areas (small parcels of them were still licensed to timber companies) to demonstrate that they were clean, that their vaccinations were up to date, and that they were free from any disease that would "pollute the water." One extreme measure even prohibited "spitting or blowing noses onto the ground and all other filthy habits," a behavioural dictum that seemed practically ludicrous considering the rugged conditions under which bush workers lived and toiled.[13] In 1925, the Greater Vancouver Water Board (GVWB) was created to oversee the orderly administration of the city's drinking water, and A. Ernest Cleveland was appointed its first chairman.

Just like in Stanley Park, however, those who sought to keep these watersheds unsullied and free from any signs of human intrusions were at the same time perfectly comfortable with manipulating these environments in the name of improving them. Although the Capilano and Seymour watersheds were praised for being sublime wilderness, during the early 1920s (and even before, but to a much smaller extent) they were fundamentally reconfigured to improve the volume of water, and the consistency of its flow, that was available to Vancouver. Tunnels were drilled to drain and redirect the flow of lakes, and dams were constructed to create large reservoirs, which flooded thousands of acres of forest and obliterated the sites of once glorious waterfalls. But such projects, however much earth they moved or rock they blasted, were never seen as detracting from or impinging upon the natural qualities of this pair of watersheds with which Vancouverites had been gifted.

For these reasons, Hopping had certainly chosen wisely when he had selected Stanley Park and Seymour Canyon – and potentially the still heavily infested Indian River watershed – as ideal sites for treatment by aerial dusting in 1930. There was still much ground to be covered, however, between pitching these projects and realizing them. Hopping knew full well that fostering broad support for them among provincial and dominion forestry officials and politicians and civic representatives would be crucial as he tried to move forward.

Hopping set about to garner the support he needed, with his efforts benefiting greatly from the remarkable results his project at the Wigwam Inn had produced the previous year. That operation had convinced the British Columbia Forest Branch (BCFB), for instance, that aerial dusting could be effective in combatting looper infestations. The same could be said of Brittingham & Young (B&Y), the owner of the Wigwam Inn and several major timber licences in the Indian River watershed. Although both the BCFB and B&Y had dragged their heels when Hopping had initially asked them to back the first dusting project in BC a year earlier, now that they had witnessed its effectiveness they were converts to his cause.

As a result, over the fall of 1929, as Hopping endeavoured to create the partnerships that he needed to realize his goals, both the BCFB and B&Y assisted him in this undertaking. For example, in mid-October Edward B. Prowd, an Assistant Forester in the office of the BCFB's Chief Forester, wrote Ralph Hopping, George Hopping's father and the dominion's senior forest entomologist in BC, to explain that he was drawing up the BCFB's budget for the following season and wanted to know about possible funding requirements for insect control work. Although the BCFB had cooperated with the dominion entomologists in the past in combatting beetle infestations in pine stands, it was unprecedented for it to offer to help pay for aerial dusting operations without the entomologists first having to come ask, caps in hand. "I understand you have been recently exploring the spread of this insect in the vicinity of Burrard Inlet," Prowd's inquiry to Hopping began. "In view of the dusting experiment carried out at Indian River, we are interested in having your opinion as to the practicability of this means of control applied on a commercial scale."[14] Ralph Hopping was unequivocal in his response. He explained that there were several major looper outbreaks in the watersheds that drained into Burrard Inlet, nearly all of which were "already epidemic," but was emphatic that "even these epidemic areas should be dusted." He thus advised Prowd to allocate $5,000 for this work in 1930, a princely sum considering the BCFB had spent merely a few hundred dollars on it the year before.[15]

Likewise, B&Y was so convinced of the need to conduct aerial dusting operations in 1930 that it facilitated a meeting of all concerned – one that reflected a truly remarkable willingness to cooperate to accomplish common goals. In early November 1929, the company leaned heavily on Peter Caverhill, BC's Chief Forester, to begin making the arrangements for these projects to proceed the following season; clearly B&Y's principals had learned the mistakes of their previously dilatory ways. Although Caverhill and his colleagues (such as Prowd) at the BCFB's headquarters believed such work was necessary and desirable, they still wanted more details about the infestations and about which ones should be addressed. They also realized the value in gathering all the interested parties in one place to develop a sound plan of action. The conference was scheduled for 19 November in the office of Charles MacFayden, the BCFB's District Forester for Vancouver.[16]

Predictably, the crux of the matter was the cost of any aerial dusting operation, and both before and after the meeting in November, Canada's leading forest entomologists involved in the effort in BC debated how to determine whether a treatment was worth applying to a particular location. While this discourse was an integral part of building a persuasive business case to justify a project's costs to senior politicos, there was an even more important element to it. The discussion reinforced the fact that North Americans generally valued different aspects of the non-human world based upon the purpose for which they used them. In this case, the merits of applying a chemical treatment to an infestation in a commercial forest could only be directly measured against the dollar value attached to the trees that were being protected.

The debate principally involved Malcolm Swaine, Canada's forest entomologist, and George Hopping, the dominion entomologist who had orchestrated the first aerial dusting operation in BC in 1929. In early November of that year, Swaine wrote to his entomologists in BC to provide the data about the cost of the projects that had been carried out against the hemlock looper back in eastern Canada. He explained that in addition to the cost of the aircraft, this insect could be killed with fifteen to twenty pounds of calcium arsenate – and at a cost of about $1.50 – per acre. Swaine then asserted that it was only practical to dust in situations where the treated areas were composed of "stands of high value" and the cost only "reasonable" if the outbreak was in its "initial stages when it is in a spotty condition," and when the spots can be "cleaned up and the outbreak prevented from spreading." Otherwise, he declared, "it would not be feasible, at least on ordinary stands to dust hundreds of square miles of infested forest." Ultimately, Swaine concluded, the final decision about whether any future aerial operations in British Columbia were warranted rested with the entomologists in that province because they alone had "an intimate knowledge of the actual ground conditions."[17]

George Hopping fundamentally disagreed with Swaine's analysis and believed that a much broader conceptualization of the subject was needed. Hopping relayed how he had often heard the old saw about control operations, whereby they "should not be out of proportion to the value of the stand protected. This statement," he argued, had "always seemed a rather empty one except in those cases where the infestation has covered vast areas and is therefore uncontrollable from a practical standpoint, the same as a forest fire which has a hundred mile front." Hopping then mounted his counter argument. "For small areas of a few square miles or even as much as ten square miles I fail to see how you can determine the value of the stand which you are protecting by a control operation. How do you know how far that infestation would have extended if not stopped?" he asked Swaine rhetorically. After recounting the millions of feet of valuable timber that individual insect

infestations had destroyed recently, Hopping pointed out that the pests would presumably continue to do so in the future if control measures were not undertaken.
"It seems to me that infestations should be looked at more in the manner of forest
fires," he asserted. "Simply because the fire happens to be burning in timber of small
value at the time is no reason for ignoring it. Nobody knows how far it may spread.
And so it seems to me," he insisted, "that the oft repeated rule about the value of
timber cannot apply except in extreme cases which are obviously hopeless, because
the amount of timber is purely problematical."[18]

Swaine remained unconvinced, and although he suggested that Hopping's case
was "interesting" and commended him, rather tongue-in-cheek, for his sudden
"burst of eloquence," he quickly launched his rebuttal. Swaine pointed out that it
rarely made sense to conduct large-scale dusting operations in eastern Canada's
pulpwood forests because of their relatively low value and the projects' relatively
high cost, but that strategic, precision strikes would sometimes be justified if they
created the potential to stamp out an infestation. In contrast, the relatively high
value of timber in BC made it possible to conduct even expensive dusting operations in that province. "Always," Swaine emphasized to Hopping, "the possible
spread of the outbreak must be taken into consideration. That is exactly the statement I gave you in my letter of November 6th."[19]

This exchange highlights the main theme that likewise framed the mid-November
1929 meeting of all the parties who had an interest in aerial dusting operations in
BC the following season, and George Hopping sought to steer the discussion in a
way that would redound to his advantage. He took charge of the gathering from
the outset, and all the representatives present – those from the GVWB, the Dominion Forest Branch (DFB), the dominion's Entomological Branch, and the BCFB –
stressed their need to know the price tag attached to any project before agreeing
to it. In an obvious effort to foster greater support for the work, Hopping stressed
that the cost of future dusting efforts could be lowered significantly – from more
than $10 to $7 per acre – by using more efficient equipment and conducting them
at the "proper time."[20] Eventually, Hopping was asked for his suggestion on how
best to proceed, and his answer was based as much on practical as it was political
considerations. He proposed conducting another cooperative project in the Indian
River watershed, but this time on a much larger control scale. Hopping was acutely
aware that treating high value, commercial timberland, which was controlled by
both the dominion and provincial governments, held out the greatest promise of
generating financial support from those governments and from the local timber
licence holder, in this case B&Y. In addition, he recommended carrying out an
experimental dusting operation in the Seymour River watershed, which was under

the auspices of the GVWB. Ultimately, all present agreed to investigate the situation further and report back within a few months.[21]

The discussion also betrayed the scepticism senior officials within the DFB continued to feel about the efficacy of this type of work. While the DFB's local forest superintendent, Ed Walmsley, remained an ardent advocate of aerial dusting, Charles MacFayden, his direct superior, and Ernest Finlayson, the DFB's director in Ottawa, both continued to argue that they could better spend their precious budgetary resources on other endeavours. Finlayson informed MacFayden on the eve of the November meeting that any funds MacFayden committed to an aerial dusting project would have to come from the "allotments" already made for his district. When MacFayden spoke during the gathering, he stressed that he would support a single project at the Indian River, that he was not committing the DFB to any expenditure at the time, and that he would pay for only half the cost of a dusting operation on dominion lands (and only to a maximum of $3,750). "I am not overly enthusiastic over this expenditure," he confided to Finlayson, "and frankly, it could not be taken from the estimates of the district."[22] Finlayson remained unconvinced of the need to take action, even after reviewing the map that clearly illustrated that the worst looper infestation in the Lower Mainland was located in the Indian River watershed where the DFB administered a large tract of valuable timber (Map 3). Walmsley's lobbying also had no effect on Finlayson. "In my opinion," Walmsley respectfully explained to the DFB's leader, "this is a very serious matter and demands immediate action if we are to combat this enemy before it has brought about a serious economic loss. It may well be that under the circumstances," Walmsley's appeal continued, "you may not feel disposed to recommend the expenditure of such a large sum of money for that purpose, but I certainly feel that the fact should be made widely known and some definite and well defined plan of cooperation, if thought advisable, arranged with the Provincial Government of this Province to counteract this enemy of the forest which threatens to destroy much of our natural forest wealth."[23] Ultimately, Walmsley's efforts were for naught, at least in 1930. Finlayson remained party to the discussions about aerial dusting that year, but the DFB chose not to participate in it.[24]

Finally, the discussions at the meeting also revealed the fundamental role that prevailing ideas and conceptions about nature played in shaping environmental policy, and the gaping chasm that could separate them from practical considerations. W.H. Powell represented the GVWB, and he made it clear that he had several compelling reasons to support any dusting control measure. He feared that the looper would lay waste to the trees in Seymour Canyon's upper reaches. Not only would the standing dead timber be prime tinder for a fire, but the defoliated forest would result in the production of huge volumes of silt and debris when it rained. Obviously, both of these scenarios could literally soil the city's supply of pristine drinking water. More importantly, however, Powell stressed another reason the GVWB was so keen to tackle this pest: he feared that the Board would be attacked for being asleep at the switch if the looper were allowed to desecrate this revered site. As Powell explained, "should the Looper show up on lower levels down the creek … public opinion will possibly subject the Board to criticism unless some advance is made against the inroads of the insect."[25]

Throughout the fall of 1929, it became increasingly apparent how George Hopping's physical distance from his entomological colleagues in central Canada compelled him to be a true pioneer in this endeavour. For example, he was gathering more information about the life cycle and behaviour of the hemlock looper in general and the western hemlock looper in particular than any North American entomologist before him. In fact, his observations and data were considered so impressive by his colleagues in the Pacific Northwest that they encouraged him to convert this information into a thesis on the subject. Some of his work would soon become crucial foundational building blocks in the entomologists' collective understanding of their field.[26] In late November 1929, for instance, he sent Swaine a dossier that was replete with information he had collected about the behaviour of the looper infestation – and the parasites that were battling it – in the Indian River watershed. One thing about the parasites stuck out to Hopping, and that was "the influence of wet weather" on them. "The more I see of these defoliators," he told Swaine, "the more it appears that the climatic factors are the most important in the rise of an epidemic."[27] In addition, Hopping was open to any idea that would further his cause. One of his men, who had been collecting data in the vicinity of the Wigwam Inn during the fall of 1929, had taken his lunch break each day in virtually the same spot outside the property's main building. In the process, he had discarded the cores of his pears in the same general patch of the forest floor, and they were soon covered in looper moths. "So many moths had gathered on them," Hopping excitedly exclaimed to Swaine, "that the cores were completely hidden. This suggests the possibility of baits."[28]

Hopping's ability and willingness to think outside the box would serve him well as the discussions about the dusting projects in BC inched forward that autumn. He was convinced that, just like Swaine back east, the operations would only be efficient and safe if they used an aircraft that was powered by three engines and boasted a payload capacity of at least 1,000 tons of poison dust. Swaine had procured just such a plane – the trimotor Ford – for his projects in northern Ontario and Quebec in 1929, and by mid-November of that year Hopping had already

begun asking Swaine whether he could use it – or another one with similar characteristics – in British Columbia the following season. If not, Hopping was hopeful
that he could "scrape up a couple of planes" on the Coast.[29] In his response, Swaine
agreed that the Ford was truly the only suitable aircraft for this type of work. In
fact, he informed Hopping that a single-engine DH61 had been used to dust in
Muskoka in 1929 and that it "was completely used up in two week's work. It is too
great a strain on any engine to run full out for so long a time, and it is necessary in
flying so low over the tree tops to run a single engine machine at its fullest power
during the time the dust is being delivered." Swaine added that Hopping might
find a commercial flying outfit that would choose to carry out these projects with
a single engine plane but he feared that the Air Board would not "allow its single
engine planes to do this work."[30]

In addressing Hopping's specific request, Swaine explained that he anticipated
undertaking at least two aerial dusting operations in eastern Canada during 1930,
which meant that the trimotor Ford "would not be available for dusting work on
the Coast."[31] As a result, just over a fortnight later Swaine reminded Hopping that
the Air Board refused to use single engine planes for dusting because the work "is
too dangerous for a machine utilizing all its power in flight. A trimotor plane is
able to carry on with two engines should one go dead and at all times has a reserve
power that may be utilized." The problem, Swaine underscored, was that there were
"very few planes in Canada other than single engine machines." He thus encouraged Hopping to locate as soon as possible a trimotor aircraft in Seattle or some-
where else on the Pacific Coast. The message was unmistakable. Hopping would
probably be on his own both in terms of finding a suitable plane and designing and
building a hopper to fit it.[32]

Those matters temporarily receded into the background as others rose to the
fore, the first of which greatly increased the chances that aerial dusting work would
occur in BC in 1930. While discussions were continuing about which sites to treat,
Swaine completed and circulated his final report on the project that had occurred
in Manicouagan, Quebec, in July 1929. That effort had targeted the hemlock looper,
the eastern cousin of the culprit that was causing such havoc in British Columbia. In
his summary of the undertaking, Swaine declared that it had been conducted "with
apparently complete success," and that the same conclusion had been drawn after
the small dusting operation had been carried out at the Wigwam Inn in BC over
roughly the same period. Swaine was unequivocal that he and his colleagues had
mastered the art of aerial dusting the insects that were harming Canada's forests.[33]

With that forceful gust of wind in his sails, Hopping changed tack slightly to
make certain that he obtained the requisite resources to treat a few sites in BC the

next season. Initially, his plan had entailed carrying out a control dusting operation in the Indian River watershed and a much smaller one at Seymour Canyon.
But few eyes would ever see the effects the application of this new technology had
on forest pests in the relatively remote Indian River, and so Hopping adjusted his
itinerary. He and his entomological colleagues had been tracking outbreaks of the
hemlock looper and hemlock tip moth (*Peronea variana*) in Stanley Park. Treating them presented him with a golden opportunity to demonstrate to all British
Columbians – and all Canadians – the importance of forest entomology in general
and the immense potential of aerial dusting to address issues involving tree pests in
particular. Moreover, practically all the other insect infestations in BC at this time
were located in mountainous terrain that presented extremely challenging conditions for aerial dusting. In sharp contrast, Stanley Park represented what Hopping
would later term "an ideal location to dust."[34] Its topography was level and surrounded by water, from which planes could easily take off and land. In addition, the
nearby False Creek was home to several flying firms, and it boasted a few fueling
stations and numerous sheds that were ideal for storing chemical dusts. Perhaps
most importantly, many roads bisected the park itself, with the main one running
nearly straight; it could serve as a great baseline for organizing the bombing runs.

So in mid-December 1929, Hopping delivered an urgent and rather sensational
appeal to the Vancouver Park Board, one that the local press heavily publicized and
that produced remarkable results. In it, he declared that it was necessary to treat
by aerial dusting all of Stanley Park's forested areas the following summer at an
estimated cost of $7,000 in order to control the insects' depredations there. To but-
tress his case, he informed the Park Board that all indications pointed to the park's
two outbreaks intensifying their destructive ways during 1930, which would cause
"very serious defoliation and stunting of growth."[35] That was all the Board needed,
particularly with Vancouver's dailies running headlines that screamed, "Aerial
Dusting Urged to Save Stanley Park."[36] It summarily approved the project, although
in the same breath it declared its intention to take the matter up with officials in
Ottawa "with a view to their co-operation and financial assistance."[37]

But sometimes the best-laid plans of mice and men go awry, often due to circumstances beyond their control, and that is precisely what seemed to be happening over the course of late 1929 and early 1930 to Hopping's designs for dusting
projects in BC. This period witnessed an unprecedented spike in unemployment,
a crash in the continent's stock markets, and desperate attempts by governments
at all levels to slash spending as the Great Depression began. Municipalities were
burdened with the lion's share of the fiscal responsibilities for providing relief to
the desperate throngs but lacked the revenue streams to do so, and so they were

squeezed particularly hard by the dynamics that were at work during these years. Vancouver became a popular gathering site for those who were down on their luck on or near the Pacific Coast, pushing the city to cut expenditures wherever it could in an effort to find the resources to support its growing ranks of needy residents.

Conducting an aerial dusting operation in Stanley Park seemed like a gratuitous expense amidst these exigent circumstances, and despite Hopping's best efforts, it soon fell by the wayside. As the province's economic situation worsened through the first few months of 1930, he sensed that his project in Vancouver's famed urban forest might be on the chopping block. He thus made some pre-emptive strikes. In early April, he wrote William Rawlings, Vancouver's Superintendent of Parks, to remind Rawlings of the pressing need to dust Stanley Park "thoroughly" because all indications pointed to the two insect outbreaks there reaching "severe proportions" that season. He simply wished to reconfirm that the city's Park Board would fund the operation.[38] George Hopping's father, Ralph, delivered the same message to his superiors in the dominion's Division of Entomology around the same time. "The importance of maintaining the timber in the park in a healthy condition," the senior Hopping declared emphatically, "cannot be overestimated."[39]

But apparently it could be, at least to the bean counters at the Vancouver Park Board. At its regular meeting in mid-April, Rawlings stressed that he was acutely aware of the need to dust the tip moth and looper outbreaks in the park. Unfortunately, however, the recent massive reduction in the Park Board's budget meant that it no longer had enough money to pay for the project. As Rawlings lamented to Hopping, the "'heavy cut' made by Council" eliminated the Board's appropriation for dusting, and he very much regretted "that the circumstances prevent us from carrying out this work, of a protective nature."[40] This news elicited a major outcry from both Hopping and Vancouver's major dailies. Hopping aggressively argued that the dusting had to be done because of the damage the pests were causing, with the tip moth in particular spoiling the "natural beauty of the trees and swarm[ing] over picnickers."[41] Moreover, the press ran editorials that urged the city to take immediate action to "save the park" and argued that the civic officials would not think twice about finding the money to save a historic building.[42]

Hopping was nothing if not resourceful, and his decision to present his case in more dramatic terms soon produced the desired result. In late April, he wrote the Park Board asking it to reconsider the matter. "The point which should be stressed is that, after the serious damage starts, it will be too late to make preparations for airplane dusting," he explained. He also invited the Board's members to visit Seymour Canyon, major parts of which had been destroyed by the hemlock looper, if they wished to see for themselves the damage that forest pests could cause.

Hopping's stern warning compelled the Board to convene a special meeting on 1 May, during which it found $7,000 in its budget to conduct the dusting campaign in Stanley Park that year.[43]

Although the Park Board had agreed to support the project, it was still mightily concerned with the optics of its decision, and so it demanded that Hopping be discreet about publicizing its about-face. Rawlings, the Board's superintendent, conveyed the good news to Hopping immediately after the Board meeting had ended. At the same time, Rawlings explained that this was a most delicate matter considering the catastrophic state of the economy in general and the city's finances in particular. As a result, he issued a blunt directive to Hopping. "It was decided," he explained, "that no further press publicity be given to this matter, so I would ask you to fall in line with the Board's wishes."[44] Hopping assured Rawlings that he understood the message loud and clear and would keep the "Stanley Park looper situation out of the press as much as possible." He also pledged that the many articles appearing recently in the local dailies had not emanated from his office. "The idea of airplane dusting is rather fantastical to the layman," he asserted, "and as a result the papers make as much as possible out of it."[45]

George Hopping's father, Ralph, then made an identical appeal to B&Y, and in doing so reminded the firm's owners of the true price tag on preserving nature's aesthetics. Reporting on the hemlock looper infestation in the Indian River watershed, within which B&Y's principals owned several major timber licences and the Wigwam Inn, the senior Hopping stressed that it was the worst outbreak in the Lower Mainland and thus an ideal candidate for a large-scale dusting operation in 1930. But in making the case, he stressed that the region's "best timber" had already been killed by the insect and thus, "from a timber value standpoint airplane dusting would not be advisable." In the same breath, however, he hit upon the nub of the matter, namely that "there is another consideration. Wigwam Inn has an approximate value of $100,000.00. If the infestation in the vicinity of the Inn is allowed to continue," he augured gloomily, "the value of the Inn as a summer resort would be practically destroyed. Under this consideration, airplane dusting would be justified."[46]

Try as the Hoppings might, however, B&Y would not be convinced that this effort was worth it. Throughout the first part of 1930, George Hopping desperately sought to confirm with his potential partners the different areas that he would treat that season. He was anxious to make all the arrangements early enough so that the caterpillars could be dusted when they were still young, an approach that would greatly reduce the expense involved. Repeatedly he contacted Joseph Grant, B&Y's legal counsel, to see if the firm would cooperate in the dusting project that season, but just as often he was rebuffed. Grant would only pepper Hopping with requests for an "exact estimate" of the project's cost and more information about the extent of the insect epidemic.[47] Grant also wanted official word from the provincial and dominion governments about how much of the bill each was willing to foot. Even when Hopping provided the data – he estimated the price tag for treating roughly two square miles at the infestation's southern end would be $9,000 – Grant remained non-committal.[48] By early May, reports showed that the parasitism in the infestation was high, but that the infestation was still very active.[49] Apparently, B&Y gambled that the former would overcome the latter, a decision that was undoubtedly rendered much easier to make because the DFB had decided that it lacked the financial resources for aerial dusting that season. As a result, the Indian River watershed fell off the docket of possible dusting sites for 1930.

Fortunately for Hopping, there was strong support for treating another hemlock looper infestation in the Lower Mainland that season. The GVWB was profoundly concerned about the epidemic that was ravaging the Seymour Canyon, one of the sources from which it drew its water supply. While the insect's activities were certainly presenting practical challenges for the GVWB, specifically in terms of creating a fire hazard and the potential for future erosion in the area, its members were just as concerned about the impact the unsightly appearance of the defoliated trees would have on those who felt that Vancouver's water was something to behold because it came from an environment that was considered pristine. Consequently, in late 1929 officials from the GVWB had responded enthusiastically to the proposal to conduct a dusting operation in the Seymour Canyon, and they maintained that position through to early May 1930. At that time, Hopping contacted Ernest Cleveland, the GVWB's commissioner, to confirm that the Water Board was still committed to the project. Cleveland unhesitatingly assured him that it was, and informed Hopping that one of the Water Board's officials was presently out in the Seymour Canyon mapping the exact area covered by the looper infestation, information that Hopping would use to draw up a work plan.[50] He informed Cleveland on 9 May that the tract the looper had infested below the dam on Seymour Creek measured a little under three square miles in size and could be dusted for a total of $12,000.[51] Hopping added reassuringly that treating it would prevent the infestation from spreading to the mouth of the waterway.[52]

Although the dusting project for the GVWB seemed like a sure thing, Hopping demonstrated an uncanny knack for sensing when a little pressure was needed to seal the deal. He was fully aware that the Water Board was subject to the same fiscal pincers that were squeezing all Canadian bureaucracies at this time, and he feared that these dynamics would cause it to suffer from the same case of cold feet that had afflicted the Vancouver Park Board with regard to its own dusting project. Hopping thus tried to forestall this situation by sending the Water Board a letter on the eve of its meeting in mid-May, the agenda for which included the proposed dusting project. "I thought," his message began, "that a few observations with regard to the Seymour Creek hemlock looper infestation might clarify the situation and help you to make a decision with regard to control measures." He then outlined how his field work in the Seymour Canyon had indicated that that spring would bring a major migration of the hemlock looper to the woods below the dam, and that it would "mean the loss of a large percentage of the timber on this area, creating a serious fire hazard. If dusting measures are undertaken this year," he confidently predicted, "there is no doubt that this timber can be saved. Airplane dusting is no longer in the experimental stage. Successful projects have been carried out both in Canada and the United States." And in case the Water Board did not wish to take his word for it, Hopping added an important coda to his appeal. "It was the opinion of my Chief," he declared without revealing that his boss was also his father (Ralph Hopping, BC's senior forest entomologist), "who inspected the Seymour area last season, that the timber below the dam should certainly be dusted and so save the timber on this area."[53]

Yet again, Hopping's adroit strike hit the mark. The next day, Ernest Cleveland, the GVWB's commissioner, profusely thanked Hopping for having delivered such a strong argument in favour of the dusting project at Seymour Creek. As much as Cleveland had supported the effort from the outset, he was acutely aware of the crucial role that Hopping's letter had played in convincing the Board that it was imperative to treat the hemlock looper outbreak that season. In mid-May, the Board confirmed its decision. Hopping was thus hard-pressed to make all the preparations needed to carry out this effort, as well as the one in Stanley Park, in the spring of 1930.[54]

In doing so, Hopping was once again forced to blaze a trail, particularly because of some news he had recently received from Ottawa. While discussions had been ongoing regarding the projects in British Columbia, Hopping had held out hope that it would be possible to borrow the trimotor Ford, which had performed effectively for Swaine while he had conducted his aerial operations in Ontario and Quebec in 1929. But this was not to be. In March 1930, Swaine explained to the entomologists in BC that there would be no large-scale dusting projects "in the east," but that the Ford would nevertheless be kept in Ottawa where experimental work would continue towards perfecting its hopper.[55] Although Swaine agreed to send all his drawings of the hopper that his men had built for the Ford to help Hopping construct one for the dusting operations in BC that season, this offer of assistance was practically worthless. Not only would Swaine's men never find the plans (apparently someone had forgotten to back up the original copies of them), but Hopping would be forced to use a fundamentally different aircraft than the one for which Swaine's hopper had been designed.[56]

The upshot saw Canada's two aerial dusting pioneers – Hopping and Swaine – debating which engineering principles to incorporate into the hopper that the former would use that season, and their exchange highlighted the truly avant-garde nature of their work. With the dusting operations in BC scheduled for early June, Hopping and Swaine were desperately bandying their design concepts back and forth. When Hopping suggested that he was going to create a tube that would inject "an air jet from the outside" and deliver a blast of wind through the mouth of the hopper, thereby eliminating the need for an agitator (which had typically broken down and jammed),[57] Swaine hurriedly wrote back to offer some important insight into the idea. In doing so, Swaine waxed philosophically about the work in which they were engaged. "There is no doubt that the disappointments are as valuable for future workers, perhaps, in some cases, as the successes," he reflected pensively to Hopping.[58]

Soon enough, Hopping had resolved his hopper and airplane hitches. Just as he had done in 1929, he contracted Western Canada Airways (WCA) to conduct the aerial dusting operations that season. Although he knew that it was optimal to carry out the work with a large and powerful aircraft that had a minimum payload of 1,000 pounds and was mounted on pontoons, none was available locally. WCA was thus forced to improvise. It would use three of its Boeing B1E Flying Boats, which were small but based at nearby Swanson Bay. Although Hopping had used this type of plane in 1929 and had helped build a hopper for it, the apparatus was in dire need of improving. Again, this challenge was rendered that much more difficult because the plane lacked pontoons and its hull rested on the water; the mouth of the hopper would thus be underwater each time the plane landed. Hopping and his colleagues worked with engineers at Boeing, fortunately based in nearby Seattle, to develop "twin hoppers," each one with a capacity of 400 pounds, that formed a saddle that fit over the plane's hull (this same design had been tried in 1929). The crew fitted them into the plane's cabin and lashed them down with cables tied to its keels.

What made this apparatus unique was that it lacked an agitator. Instead, to prevent the dust from packing, "a tilting floor was arranged on the slopes of the hopper."[59] As the dust spewed out its door, heavy springs raised the hopper's floor so that as the volume of dust in the hopper decreased, the slope of its floor increased. In this way, a constant flow of poison dust was maintained.

6.1 Loading hopper with dust for Stanley Park project. Source: Natural Resources Canada, Pacific Forestry Centre, IndianRiver001.

Undeniably, Hopping had faced huge challenges in terms of his equipment, but he proved to be an adept innovator.[60]

In other ways, Hopping benefited from his symbiotic relationship with Swaine and the latter's advice and experience, particularly the reams of data the latter had gathered from several seasons of aerial dusting back east. For example, Hopping had organized and conducted the project at the Wigwam Inn in 1929 on such short notice and had treated such a small area that he had not been able to determine how far apart the plane's passes should be in order to achieve the desired distribution of poison dust; Swaine gladly sent Hopping all the figures he had on this subject. Likewise, Swaine informed Hopping that he had perfected the art of reloading the plane with dust during the previous season. Instead of the plane taxiing to the shore between each flight, Swaine's men had transported the bags of dust out to the plane by boat and then formed an assembly line to pass the bags of dust from the scow and dump them into the plane's hopper. Using this method, Swaine boasted to Hopping, the former had been able to load 1,200 pounds of dust in three minutes (this process had sometimes previously taken as long as a half hour). By mid-June, Swaine had

6.2 The scow that was used to load the plane with dust for the Stanley Park project. Source: Natural Resources Canada, Pacific Forestry Centre, IndianRiver003.

reviewed all the preparations that Hopping had made for the dusting projects in BC that year, and he commended Hopping for his efforts. Although Swaine cautioned Hopping not to bite off more than he could chew – Swaine was under pressure to pare expenses and aerial dusting work carried a considerable price tag – Swaine lauded Hopping for having made "careful plans" and delivered his "best wishes for success."[61]

Throughout the latter part of the spring, Hopping was consumed with addressing the projects' remaining practical details. He met with officials from the BCFB to review the dusting missions planned for that season. He did the same with representatives from WCA, and stuck with this contractor even though other flying firms lobbied for these jobs. Hopping informed WCA that there were about 200 acres of Stanley Park that its pilots would not be treating, specifically the park's open areas (largely playgrounds and athletic fields) where the trees were so widely spaced that it would be prohibitively expensive to dust them from the air. Instead, ground sprayers would be used in these areas to shower the trees with poison.[62] Hopping also ordered the calcium arsenate – 28,000 pounds in all and delivered in fifty-pound bags – and had it stored in a base out of which WCA operated in nearby False Creek.[63] With the press predicting in early June that the dusting operations in Stanley Park would soon begin, all that Hopping needed was favourable weather.[64]

It came in the middle of the month, when Hopping's preparations began paying off handsomely. Roughly 800 of Stanley Park's 1,000 acres were slated for aerial treatment, and the targeted stands were neatly divided into small pieces by the park's many trails and roads. Not only did the latter serve as invaluable guides for the pilots who conducted the dusting missions, but they also allowed Hopping and his men to measure the size of the individual plots and thus calculate fairly precisely the volume of dust needed to treat each one. On the night of 14 June, the scow was loaded with the requisite fuel and calcium arsenate poison dust (16,000 pounds worth), and at three o'clock the next morning it was towed into Vancouver Harbour a short distance from the park's shoreline. As Hopping's subsequent report explained, "there was just room enough between it and the park for a good take-off and landing. The distance flown to and from the loading scow was thus minimized." The three aircraft flew in rotation to facilitate loading and refuelling, and Hopping communicated with the pilots – it was still rare for planes to have radios at this time – by using a "fast launch" that travelled between the park's main entrance and the scow. The work began at four o'clock in the morning, with the heavy dew of that early hour helping the dust adhere to the trees' foliage. Within six hours, the trio of planes had bombed Stanley Park with eight tons of poison dust, a feat that Hopping noted was quite remarkable considering the challenge the pilots faced in dodging "the number of spike-top cedars, which extend far above the general forest level." Hopping noted how "in some instances the pilots flew below these, skilfully avoiding them." He had thus come to appreciate the value of using smaller planes for this type of work, because exercising this much dexterity "probably would not have been possible with larger machines." In total, the operation cost just under $6,400, or roughly $8 per acre.[65]

Four days later, it was time for the operation in Seymour Canyon, but the local geography and topography made it a much more complicated mission. Again, the scow was loaded with the necessary supplies of fuel and dust, and it was anchored near the mouth of the Seymour River. But the area slated for dusting was located eight miles inland, making a round trip to the drop zone – including turns and "reconnoitring" – roughly twenty miles in total (Map 3). Moreover, the gulch was narrow with steep walls, making it much more difficult for the pilots to line up their dusting passes. As a result, it took twelve hours to drop the same amount of dust (16,000 pounds) that had been drifted over Stanley Park in half the time, and the planes ended up flying nearly 700 miles. The project had also taken practically the entire day because one of the hoppers had broken early in the morning and had to be repaired before the dusting could continue. Nevertheless, when the costs were tallied, the price tag on this project ended up being cheaper than the one for the park because WCA had charged Hopping a flat rate for the job. The total bill came to roughly $5,900, or about $7.33 per acre.[66]

6.3 Dusting Seymour Canyon. Source: Natural Resources Canada, Pacific Forestry Centre, IndianRiver005.

Not only had Hopping's assiduous planning made these aerial dusting projects highly efficient, but the reports on their effectiveness were replete with good news. Although he openly admitted that he and his colleagues still lacked a dependable method for calculating the exact degree of mortality among the insect populations they targeted, the heaps of dead and dying caterpillars in both Stanley Park and in the part of Seymour Canyon that had been treated were a convincing testament of the degree to which the calcium arsenate dust killed the hemlock looper and hemlock tip moth. "The success of these two operations was gratifying to all concerned," Hopping triumphantly declared. "It was estimated that in the case of Stanley Park from 75% to 85% of the larvae were destroyed, and on the Seymour Area, from 80% to 85%."[67] It must have pleased Hopping and the Park Board mightily when he conducted further investigations in Stanley Park in the spring of 1931 and concluded that no more treatments would be needed.[68]

After Hopping examined both sites over the course of July and August, and then again in the first half of 1931, he was able to convey the positive results even more

6.4 Boeing B1E Flying Boat dusting Seymour Canyon. Source: University of Winnipeg Archives, Western Canada Pictorial Index, James A. Richardson Collection, Image 3460.

authoritatively. For example, he informed Ernest Cleveland, the Chairman of the GVWB, that the dusting operation on the Seymour Watershed was an unqualified success: "I think that now we can definitely say that the stand which was dusted is definitely out of danger."[69] For his part, Cleveland had, on the GVWB's behalf, already thanked Hopping for having taken steps "to preserve this timber."[70] Likewise, Hopping wrote the Vancouver Park Board and the BCFB to inform them that the dusting work in both Stanley Park and Seymour Canyon had collectively been "a complete success."[71]

But Hopping also realized during his reconnaissance trips through the treated sites that he could not take all the credit for the projects' success; nature had given them – and continued to give them – a helping hand. In his reports on the dusting operations in 1930, he identified the major role that parasitism was playing in knocking back the infestations in both Stanley Park and Seymour Canyon. He predicted that any of their vestiges would be wiped out by two parasites, *Winthemia cilitibia* (Rand) and a small *proctotrypid*, both of which feasted on the looper's eggs. He also openly admitted that, with regard to the *cilitibia*, if the infestations had been allowed to run their course, "the parasitism from this one species should have reached 50% next year."[72] Moreover, he confided to William Rawlings, the Park Board's superintendent, that he had hedged his bets. Soon after the dusting had occurred, Hopping had seen to it that four hundred parasites were released in Stanley Park, explaining to Rawlings that they should be able to gain "a foothold and cope with any loopers that appear in the next few years."[73]

Nevertheless, lest anyone suggest that this parasitic factor had rendered that season's dusting operations redundant, both George and Ralph Hopping and their boss back in Ottawa, Swaine, were at the ready to rebut such charges by emphasizing, as Swaine put it, "the peculiar value of these two areas."[74] In a report on entomological issues in BC in 1930 that George Hopping sent Swaine in Ottawa, the former tackled this issue head on. "Although the parasitism may have handled the larvae on the two areas in question," he began, "the feeding undoubtedly would have been much more severe, and many hemlocks would have been killed, spoiling the appearance of Stanley Park and creating a serious fire hazard on that part of the Seymour watershed through which the large conduit runs. Thus the dusting could not be considered anything but successful, both from the standpoint of results and the increased knowledge of dusting methods."[75] George's father Ralph Hopping hammered these points home in subsequent summaries he provided on the dusting experiments in BC that season. Referring to the project in Stanley Park and whether parasitism would have knocked back the infestation there on its own, he declared that "on account of the very high value of this recreational area, no chance could be taken, and the dusting of it was imperative." Finally, he reiterated that airplane dusting in BC would rarely be cost effective, "except where the stand has some value other than commercial, such as in a public park or on some estate."[76]

Even though Swaine had played virtually no role in the projects in BC that season, he was determined to exploit their success to foster broader support for aerial dusting work across the entire country. George Hopping had sent to him a preliminary report on the efforts in Stanley Park and at Seymour Creek, and Swaine was delighted by both its form and content. Complimenting Hopping for "the fine piece of work and a good report," Swaine explained that he planned on forwarding "the gist of [his] report over to the Deputy Minister's office."[77]

Swaine's comments capture the cooperative élan that drove Canada's pioneering entomology program during this period, the same spirit that helped its participants maximize the knowledge they gained. Although thousands of miles separated Hopping from Swaine, they linked arms – figuratively – whenever they could. This was abundantly clear from their work with the hemlock looper's parasites, and a legion of others. For some time, both George Hopping and his father, Ralph, had been sending looper parasites back east to Swaine and to other entomological officials to be analysed and reproduced. These shipments were soon redirected to the new Dominion Parasite Laboratory in Belleville, Ontario, a facility that opened in 1929 and greatly enhanced Canada's ability to breed insects deemed helpful in fighting food and forest pests. The two Hoppings had also been dispatching hemlock loopers that had become infested with a particular fungus to H.T. Gussow, the Dominion Botanist, to learn more about this organism.[78]

Likewise, George Hopping's most recent dusting projects in BC had taught him invaluable lessons about how best to exploit nature's own power to battle the elements humans found undesirable within it. Just as those who battled blazes in the woods were learning how to fight fire with fire, Hopping was greatly expanding his understanding of how to fight pests using pests, and he eagerly shared this knowledge with Swaine and all the country's forest entomologists. While Hopping had identified over a dozen parasites of the hemlock looper, arguably his greatest breakthrough occurred in his study of its most "effective larval parasite," the *Winthemia cilitibia*, specifically when it laid its eggs. He had observed that it emerged as an adult typically between 20 June and 20 July, and that it laid most of its eggs on the host looper after the latter's third moult (beginning around 7 July). He stressed that this revelation had major implications for the broader campaign against the hemlock looper in particular, and all forest pests in general.

Aerial dusting was now certainly one instrument in the entomologists' tool box, but as Hopping's experience in 1930 had highlighted, it only made sense to exploit simultaneously the mechanisms that nature also provided – specifically parasites – to combat undesirable insects in the woods. To realize this end, Hopping stressed the need to synchronize aerial dusting with both the development of the target insects *and* those bugs that feasted on them. In terms of the hemlock looper, for instance, if the dusting occurred after the parasite had laid most of its eggs on the host insect (i.e., 7 July), then it would also wipe out many of the parasites alongside the looper larvae. This made dusting earlier in the looper's life cycle doubly attractive, he stressed, for this allowed for the target insect to be "killed more easily," and for the Tachynid parasite to later deposit its eggs on those larvae that escaped without "the natural increase of the parasite [being] disturbed except by the limitation of the food supply." In addition, Hopping's research had strongly suggested that there was a direct corelation between the prevalence of this parasite and hemlock looper outbreaks. As a result, he posited that "the percentage of parasitism of this species may prove to be a reliable index of the trend of looper epidemics in British Columbia and will probably aid in the advisability of control measures."[79]

In the meantime, Hopping was able to savour his recent accomplishments, particularly because those involved in such undertakings were acutely aware of the degree to which he was breaking new ground in their field. In essence, Hopping became *the* regional authority – the go-to source – for information on aerial dusting in general and employing this technology against the hemlock looper in particular. It was highly predictable that officials from both the Vancouver Park Board and the GVWB praised him for his efforts, but his projects' success caused word of them to spread near and far. For example, Franklin P. Keen was a senior forest entomologist with the United States Department of Agriculture who had been working in the Pacific Northwest since the mid-1910s. In the fall of 1930, Keen wrote Hopping to ask for details about the dusting projects that the latter had overseen that season, having been prompted to do so by news of a major hemlock looper outbreak at the mouth of the Columbia River in Washington State. He informed Hopping that government and industry officials were contemplating an aerial dusting operation in 1931 to combat the infestation, but before making a decision they wished to gather as much information as possible about the work Hopping had overseen in BC. As part of this effort, Keen had also asked Swaine for details regarding the new hopper the latter had recently designed, a model that Swaine believed would be "quite superior for this work." In writing to Hopping, Keen clarified why he was not asking for the same information from Hopping: "As I remember, you explained that the hopper which you used was more or less a home-made affair."[80]

Thereafter, Hopping's foreign admirers continued to pepper him with questions about the work he had done in BC. An article he authored on the subject appeared in January 1931 in *The Timberman*, a leading American forest industry journal, but it apparently did not provide the degree of detail sought by those who were eager to follow in Hopping's footsteps.[81] Charles S. Cowan, formerly with the BCFB but presently the Chief Fire Warden for the Washington Forest Fire Association (now the Washington Forest Protection Association), asked for Hopping's counsel on the eve of conducting an aerial dusting campaign against the hemlock looper outbreak at the mouth of the Columbia River that spring.[82] Although Cowan also sent inquiries to the BCFB and the GVWB, everyone recommended contacting Hopping because he was the virtuoso in the field.[83]

But Hopping's extraordinary success in these dusting trials in British Columbia in no way solved the province's long-term pest problems as they pertained to trees. Even after the projects in Stanley Park and Seymour Canyon had been completed in June 1930, insects continued to plague many residents in BC. Some of them, such as Members of the Legislative Assembly, were once again grappling with bug issues literally under their feet. A senior official with the BCFB in Victoria sent

Ralph Hopping, George Hopping's father, a specimen taken from the hordes that were chomping their way through the cedar timbers that formed the foundation for the BC Legislature and asked for advice in dealing with them. The senior Hopping identified the culprits as termites and suggested using a solution of carbon bisulphide to destroy them. In the same breath, however, he warned that this elixir was highly flammable and thus may create a bigger problem than the one that they were trying to solve![84]

For several reasons, BC's bug problems not only continued but intensified. Industrial exploitation pushed ever deeper into the province's forests, creating disturbances and post-harvest conditions that were often conducive to fostering insect epidemics. And the more the entomologists learned about the insects that sought to destroy the province's trees, the more they realized the true extent of the issue and how much remained to be learned.[85]

Moreover, everyone involved in BC's forestry sector would endure some growing pains as they sought to tackle the province's pest problems, particularly because stories of Hopping's successful dusting operations over 1929–30 created often unrealistic expectations about the ability of forest entomologists to resolve them. An incident in mid-1931 illustrates this well. At that time, the Campbell River Timber Company had requested that Hopping investigate the situation on one of the firm's harvest blocks because many of its Douglas fir trees were dying. Hopping did so in early July and produced a report on his findings. It identified that the main culprits were bark beetles, and he recommended a half-dozen measures that the company could take to deal with the pests. R.C. Richardson, the firm's general manager, received the report and cut it to shreds. He excoriated Hopping for having carried out a "very cursory" investigation, and for having provided "no definite practical recommendation as to the feasibility of eradicating the pest."[86] Hopping was rightly insulted by the attack and responded in kind. He stressed that he had conducted a thorough review of the situation and had recommended that Richardson undertake a handful of measures to mitigate the problem. "I am sure I do not know what more Mr. Richardson could expect," Hopping confided to a senior forestry official with the BCFB. "We have told him what was wrong with his timber and what he can do about it. If he expected a detailed cruise to determine how much of his timber has been killed," Hopping stressed, "I should advise him that to do such is not the function of our service."[87] At this point Richardson tried to make amends. He apologized profusely to Hopping for his earlier letter, stating that he had never intended to create the impression that Hopping had not done his job properly. The problem, Richardson argued, was that he was merely a rookie in the campaign against forest pests. "This is so new to me," he stressed to Hopping, "that

I probably did not realize the full import of your report which I read so quickly and passed on to my principals."[88]

Although Hopping's aerial projects in BC over the course of 1929–30 have been largely forgotten, they left an indelible impression upon those who were involved at the time. In the late 1930s, for instance, an industry official contacted Chauncey D. Orchard, BC's Assistant Chief Forester, to inquire about the hemlock looper that was menacing the firm's timber limits.[89] In his response, Orchard explained that this insect was a native and was always present in the forest, and although its population fluctuated, it was kept "pretty well in check by its natural enemies." He also mentioned, however, that experiments had been conducted wherein calcium arsenate had been dropped by aircraft to combat looper infestations. The fact that Orchard declared that the latter means had been established as a "fairly effective control measure" spoke to the short-term legacy of Hopping's cutting edge and highly successful campaign to employ this rudimentary technology to combat pests in BC's forests.[90]

"We feel that the technique of airplane dusting has now been perfected"

OUR ENIGMATIC VIEW OF NATURE AND THE LESSONS TO BE DRAWN

The interwar years were monumental for Canadian forest entomology, but they did not start out that way. Malcolm Swaine, who had been serving as the dominion government's first forest insect specialist since the early 1910s, was acutely aware of the futility of much of his work during the nascent stages of his career. As much as he could study the pests that were ravaging the country's woodlands, and as proficient as he was at conducting such investigations, he knew that he and his colleagues were practically powerless to do anything about them. There was simply no practical way to combat insects that attacked trees that were relatively inaccessible in the country's wilderness. Moreover, even if Swaine and his colleagues could have reached a stand of hemlock or balsam fir that was under siege, it was practically impossible for them to scale the height of the trees in order to parry the menace that the bugs represented.

All that changed, however, with the advent of aerial dusting, so much so that by the first years of the Great Depression, Swaine could barely contain his enthusiasm for the boon that it represented to his work. Beginning in the mid-1920s, Swaine spearheaded the effort to apply this technology to Canada's forests in an attempt to protect them from a variety of pests, and his campaign was extraordinarily effective. The dusting operations he had supported against the spruce budworm in eastern Nova Scotia and northern Ontario (north of Sudbury) had certainly produced mixed results at best, but the other projects, which had occurred in central Ontario (in Muskoka), Quebec (in the Manicouagan River valley), and British Columbia (at the Wigwam Inn, Stanley Park, and Seymour Canyon) and had targeted the hemlock looper, had been a smashing success. In reflecting upon his achievements in

1929, his gleeful remarks spoke to his conviction that he was on the cusp of a major breakthrough. "We feel that the technique of airplane dusting," Swaine declared at the time, "has now been perfected to the degree that we can feel confident placing the required dose on a definite area."[1] In his eyes, his small cadre of dedicated entomologists had breached the divide between success and failure in this type of work. "It may be said that in airplane dusting we have at last a weapon with which at least some of the destructive insect outbreaks of the forest may be effectively combatted," Swaine broadcast spiritedly in 1930. In the same breath he prophesied that "if improvements in poisoned dusts and the developments of air machines specially adapted to this purpose keep pace with the investigations, this method of control should have a wide field of usefulness in the future."[2]

What made Swaine and his colleagues' accomplishments so remarkable was that this tiny handful of pioneering scientists had established Canada as a world leader in this field. The foregoing pages have documented how interested observers near and far saw the work done by Swaine and his fellow entomologists as reaching unprecedented heights in the aerial war against forest insects. Not only did the leading international forest industry and aviation journals heap plaudits upon them for conducting their projects so effectively, but this coverage was undoubtedly responsible for eliciting at least some of the many inquiries they received from all corners of the globe asking for advice about how to carry out such efforts. In addition to the examples already cited in the preceding chapters, the postmarks on the more noteworthy congratulatory missives spoke to the global recognition the Canadians were receiving at the time. One of the most intriguing was a request for information that arrived from the State Research Institute of Agricultural Aviation in Moscow. The telegram explained that the Soviets were now conducting their own aerial dusting experiments to control forest and farm pests, and they had heard that "such works were conducted in Canada, and that the reports about them were published by your Department." A debate ensued about the potential security issues the sharing of such information posed, but ultimately officials in Ottawa complied with the Soviets' request.[3] Officials in Britain had also intermittently sought the latest updates about what the Canadians were doing in terms of bombing woodland pests with poisons, and in 1934 a special request for information arrived on Swaine's desk. George S. Vanier, Secretary to the High Commission in Britain, had sent it after entomologists at the Royal College of Sciences in London had asked him for published papers, photos, and lantern slides pertaining to the Canadians' cutting-edge work.[4]

Above all others, American entomologists exalted their Canadian colleagues for the latter's leadership in aerial dusting forest insects, and they continually solicited their northern neighbours' insight into how best to conduct these operations. Over the course of 1930–1, for example, forestry and entomology officials in Washington State contacted both Swaine and George Hopping, who had orchestrated the projects in British Columbia, about possible strategies for tackling a hemlock looper infestation south of the border at the mouth of the Columbia River. In forwarding his queries, Franklin P. Keen, a veteran forest entomologist with the United States Department of Agriculture, repeatedly stressed that the Canadians were truly *the* authorities on aerial dusting over woodlands.[5] Although Keen was anxious to collect as much data as he could on the subject, he exercised remarkable deference to Hopping in particular to ensure that Hopping received due credit for all the groundbreaking work Hopping had done. By this time, Hopping had taken a leave of absence from the Canadian Entomological Branch in order to pursue his MSc at Iowa State College. He rightly predicted that he would complete his degree quickly because, as he excitedly told one of his professors, by the time he entered his program his "thesis practically will be completed."[6] Based upon the research he had carried out over the previous few years in BC, Hopping's thesis documented the hemlock looper's life history and his own efforts to combat this pest through aerial dusting. Hopping gladly wrote Keen to share his knowledge of the subject and proposed sending Keen a draft of his thesis for the latter to peruse. "I certainly appreciate your very generous offer to let me study your manuscript," Keen told Hopping in respectfully declining it, "but I feel that this would be asking too much of you. I will probably have to write some articles on our present outbreak, and if I saw your material, I should never know to what extent I had unintentionally plagiarized your data. Better wait until it is published, then I shall be most anxious to read all that you have learned about the insect."[7]

Despite all the accolades, Swaine was determined to improve his team's capacity to use planes to dust forest insects with poisons, but his designs fell victim to much larger forces that soon subsumed them. Over the course of late 1929, for instance, he had plotted a strategy for improving the entomologists' campaign against insects that fed by burrowing into a tree's foliage by conducting experiments against the budworm and jack pine sawfly in northern Ontario.[8] The Department of National Defence (DND) actually agreed to fly these trials the next season, but they would never get off the ground. The crash of the New York Stock Exchange in the fall of 1929 touched off nearly a decade of tough economic times for the dominion bureaucracy, including its entomology branch. As a result, it was able to carry out only a few relatively small-scale initiatives, such as developing a better hopper and mapping an infestation of a new forest insect – the imported spruce sawfly – that first appeared in the early 1930s in the Gaspé area of Quebec.[9] The dominion hosted no more major aerial operations to combat forest insects until well after the Great Depression had ended, although Canadian forest entomologists continued to make progress in other realms during the 1930s.[10]

Nevertheless, analysing the projects that Canada conducted between 1927 and 1930 highlights a few important threads that weave their way through this story. In addition to highlighting the degree to which Canada became a world leader in carrying out this type of work, retelling the tale of these aerial dusting projects also underscores the way in which Americans have historically influenced Canada's natural resource policy. Much of the work that Swaine and his colleagues did to kill forest insects during this period was done at the behest of persons or firms that were based in the United States. Similarly, this story also reminds us of the preponderant influence that class plays in determining environmental policy. Time and time again, it was the elite – politicians and their small coterie of powerful allies – who determined when, where, and how these aerial dusting efforts occurred. This was all about inside lobbying. And whereas in other realms scientists have so often had their agendas shaped by outside forces, the forest entomologists and foresters in this story enjoyed a generally symbiotic relationship with the political insiders who brought their powers of persuasion to bear. The former wanted to experiment with aerial dusting and were relatively unconcerned about where they did so; the latter were eager to pull the political strings to realize that goal, as long as the efforts were undertaken to their liking. This was truly a conjunction of interests.

Other themes wind their way through this story as well. One is the exceptional degree of cooperation among different levels and branches of government and their industrial partners. This aspect of the enterprise was so noteworthy because all of the partners involved were able to act in concert without signing contracts or agreements that spelled out their respective obligations and duties. All that they needed to work together effectively were handshakes and verbal commitments, and this synergy continued to dominate forest entomology work in Canada for years to come.[11]

More importantly, nature – or the non-human environment – was the axis around which this story revolved, and it did so on two basic planes. In one sense, it exerted a considerable influence on humans. The entomologists who studied forest insects came to appreciate the ecology of the woods in which these bugs lived and how human activities were affecting the pests' populations, and often not in a good way. Logging was generally creating an ideal environment in which forest insects could thrive, and the scientists thus deduced that the best way to respond to nature in this instance was to change the way humans harvested the woodlands. In the short term, however, the only possible response to an infestation was to experiment with aerial dusting and other control measures. The entomologists also learned how simple differences in the life cycles of insects – that of the spruce budworm compared to that of the hemlock looper, for instance – had an enormous impact on the pests' vulnerability to the aerial campaigns that were intended to kill them. The fact that the spruce budworm was protected from the poison dusts while ensconced within the tree's bud, for example, while the looper was not, dictated that Swaine and his colleagues would have to employ different tactics in targeting these two forest pests.

Nature also served as the setting for this story, and as a result, human conceptions and ideas about the environment fundamentally shaped how events played out; this is, in fact, the pith of this tale. The preceding pages have described in detail the unique lens through which humans in the early twentieth century experienced the woods. On the one hand, a large cast of environmental preservationists and recreationists professed their love for their trees because their wooded retreats represented some of the most beautiful features that the natural world had to offer. Whether it be the eastern hemlock that lined the shores of Muskoka's most famous lakes or the mammoth western red cedar and Douglas fir in and around Stanley Park, the Wigwam Inn, and Seymour Canyon in British Columbia, Canada's trees had human observers worshipping at their wooden altars. As a result, these individuals zealously fought to take whatever measures they could to maintain these trees in their "natural state."

In doing so, the tree lovers' behaviour was most perplexing. Once their hemlock or fir were subject to attack by a forest insect, all agreed that the trees' transformation from having a healthy to sickly appearance had rendered them eyesores. They thus demanded that the authorities carpet the besieged stands in a toxic dust. The hope in doing so was to eradicate the pest from the local ecological equation and arrest the damage it was causing. In other words, only by keeping the trees in a state they considered lovely would their prized wilderness site retain its value. Ugly trees were simply worthless. Seen through this highly selective lens, those who professed a deep affection for nature considered certain parts of it, like the trees, worthy of protection, and they demanded that the government kill the part of it that they despised, namely the bugs. At the same time, they were blind to the many unmistakable signs of human influence in the same woods that they ironically treasured for its pristine qualities. They viewed the forest as a healing environment, a sacred place in which one could repose from the hectic and taxing grind of daily urban life, and they viewed it in this light only because they perceived it as being untouched by human hands. Nevertheless, Muskoka's seasonal residents and visitors to the Wigwam Inn, for example, accepted their numerous and often impressive summer homes, smoke-spewing steamers, and gas-burning motorboats as benign parts of the local environment, but the thought of a native forest insect devouring some of their trees was an abominable notion they could not countenance. Only by perceiving their seasonal retreats from this perspective could they rationalize poisoning the very thing – nature – that had drawn them to wilderness in the first place.

There were also other, equally ironic peculiarities in the conceptions of the non-human world that prevailed at this time (and, it could be argued, still prevail). One is best illustrated by the circumstances surrounding the aerial dusting effort in Muskoka (see chapter 4). There, the local recreationists had insisted that the government take urgent and drastic measures to save their arboreal splendour from the loopers' ravages because they considered the trees to be beautiful only as long as the hemlock remained verdant and healthy. But their case was disingenuous, because dead trees had always been an integral part of the woodlands that Muskoka's seasonal residents cherished. This reality is best illustrated – literally – in the pamphlets and brochures that publicized Muskoka's environmental attributes to potential tourists at the time (and for decades before and after). Predictably, trees are prominent in the illustrations that were included in these promotional materials, but there is a most curious aspect to them. Indeed, many of them appear as stereotypical healthy conifers whose vitality is unmistakable. However, a surprising number of them are unequivocally dead or dying. Their lifelessness is instantly recognizable in their scraggly and skeletal crowns and branches, which are void of any signs of vigour. Undoubtedly, local crows and raptors appreciated these chicots, as the lumbermen called them, for the roosts they provided. Apparently so, too, did the local tourist boosters, who felt that their inclusion in publicity booklets about Muskoka would enhance its allure. Typical are these turn of the twentieth century images from tourist brochures that illustrated Muskoka's natural charm and in which dead trees are unmistakable. The same is true of the picture that accompanied a book that broadcast the region's wooded lustre to the world. Significantly, dead trees are conspicuous in both photographs and hand-drawn images; photographers and artists alike clearly considered them to be objects that helped publicize Muskoka's unmatched vistas and thus worthy of inclusion in their pieces.

So the heart of the matter was not whether a tree was dead but how it died, when this happened, and whether anyone observed its demise. Evidently in the case of the local cottagers who demanded that the government dump tons of toxins on their hemlock, they could not tolerate bearing witness to their stands of trees perishing. Paradoxically, however, they presumably would have embraced the dead trees as being integral to the beautiful local landscape if they had simply arrived

C.1 Advertisement illustrating the beauty in Muskoka's alive and dead trees, ca. 1901. Source: Grand Trunk Railway Company of Canada and Muskoka Navigation Company, *Highlands of Ontario – Muskoka Lakes* (Battle Creek, MI: Gage Printing, 1901), cover.

A CHARMING BIT OF LAKE ROSSEAU NEAR "ROYAL MUSKOKA" HOTEL.

C.2 Advertisement illustrating the beauty in Muskoka's alive and dead trees, ca. 1905.
Source: *Royal Muskoka, Muskoka Lakes, Canada* (Gravenhurst, ON: Muskoka Lakes Navigation
and Hotel Company, 1905).

C.3 Promotional image illustrating the beauty in Muskoka's alive and dead trees, ca. 1879. Note the
prominence of the dead trees in this print that was included in a tourist brochure about Muskoka.
Source: J. Rogers, *Guide Book and Atlas of Muskoka and Parry Sound* (Toronto: H.R. Page, 1879), 27.

on the scene once the hemlock looper had done its business. This absurd situation raises the proverbial question: if a tree died and no one watched its demise, did it maintain its charm?

There was another bizarre angle to the love of nature that humans expressed in this story, and it was reinforced by the elitism that dictated where the aerial dusting projects took place. Apparently all beautiful trees were not created equal, even when they were the same species and appreciated in the same way for their aesthetic qualities. What really mattered was whose trees they were and where they were located. Around the same time that Muskoka's seasonal residents were conveying their concerns about the local forest pest infestation in 1927, the identical issue was eliciting equally compelling cries for help a few hundred miles away to the southeast in the vicinity of Brockville, a locale with a very different demographic (Map 1). Located on the north shore of the St. Lawrence River, this city was a significant industrial and commercial centre within what was a relatively poorer area of the province. Whereas the folks who outlined their deep anxieties about the insect epidemic in Muskoka represented society's upper crust, Brockville's residents generally occupied the middle of the socio-economic spectrum. In 1927, Brockville's citizens had noticed that many of the conifers growing just back from the river's banks for a distance of roughly fifteen miles southeast of town had begun turning brown and losing their needles. Margaret Cornell, a local from Brockville, was so chagrined by this that she was spurred to convey her concerns to the Canadian Department of Agriculture in August of that year. "Lovers of trees in this vicinity are dismayed by the rapid destruction of hemlock and other evergreen trees," she explained. She described how, roughly four miles east of Brockville, thousands of trees had been stricken within the last few weeks and now appeared to be entirely brown and dead. "They are festooned with the cobbwebby refuse of the so-called 'inch-worm,'" Cornell declared. To buttress her case, she included part of a tree's stripped branch with a living worm on it with her letter. She added that she had penned her missive on her own behalf, but also on behalf of many others, all of whom were most anxious to "know if anything can be done to stop the destruction. Also, is there a possibility of these trees living?" Provincial and dominion officials soon confirmed that the culprit was the hemlock looper.[12]

What occurred next convincingly illustrates the enormous role that class and power play in shaping environmental policy, and the capricious way in which humans value nature. It will be recalled that Muskoka's hemlock were considered priceless because of their aesthetic value. As a result, when the well-heeled recreationists there demanded that the government take immediate steps to combat the pests that were attacking these trees, public officials took extreme measures and spent thousands of taxpayers' dollars to address the issue. In sharp contrast, however, the same public officials elected not to take any action to tackle the looper menace in the Brockville area, even though the locals there praised their exquisite hemlock in terms that were just as emotive as those used by Muskoka's seasonal residents. Significantly, the dominion entomologist who answered queries regarding the looper outbreak near Brockville informed concerned locals that airplane dusting represented a possible solution to the problem but that it "involves great costs and is not without its danger."[13] Apparently these considerations were insurmountable hurdles to bombing the pests with poison in the St. Lawrence River valley, whereas the provincial government quickly found the money and risked several pilots' lives to carry out the aerial dusting work in Muskoka. As a result, within short order the damage had been done in eastern Ontario. As the dominion entomologists reported in late 1928, "in the Brockville district where the infestation was known to have occurred last year, all hemlock trees are dead and the insect has disappeared."[14]

To be sure, it was not strictly the recreationists who viewed nature through such a peculiar prism. Nearly all the participants in these aerial dusting trials were blind to the collateral damage that spreading their poison dusts was causing. In reviewing the surviving documents from these projects, it was shockingly rare that someone – anyone – raised concerns about the deleterious impact that releasing tons of deadly chemicals over woodlands would have on non-targeted organisms, including the humans who actually carried out the work. This indifference was all the more significant because the evidence of the ancillary damage kept appearing right before their very eyes. It was also noteworthy because it differed fundamentally from the situation in the United States. There, officials involved in using chemical insecticides prior to the Second World War knowingly downplayed or dismissed the toxins' dangers because of their vested interest in fostering support for their dusting programs. In contrast, the decision by those involved in the Canadian forest dusting efforts during the interwar years to ignore these hazards was not motivated by any such conspiratorial or pecuniary force.[15] Swaine and his fellow scientists undoubtedly knew that arsenate dusts were lethal to humans and other living things. Nevertheless, they had such a disjointed view of nature that they seemingly could not fathom that the same toxins that could kill forests pests would be at least harmful if not fatal to humans and animals of all sorts, and poisonous to supplies of drinking water.[16]

The Canadian entomologists had certainly been given sufficient advance warning about these potential problems, but they either chose to ignore such warnings or were deaf to them. Astoundingly, Swaine himself had been one of the first messengers. Back in 1924, he and Frank Craighead had published their landmark study of the spruce budworm. In it, they had predicted that it might be possible in the future to use "artificial methods of control" – such as aerial dusting – to contain outbreaks of this most destructive forest pest. In the same breath, however, they openly wondered whether "the effects of wholesale application of poison in the forest might be very injurious to bird and animal life."[17] Moreover, when the Canadians had sought out and received reports of the experimental aerial dusting work that the French had carried out in the mid-1920s, the documents that they received from overseas were peppered with safety warnings. The French had explained that they had chosen not to dust with either lead arsenate or sodium arsenate because of their toxicity to humans (although the Canadians used mostly calcium arsenate, they did drop a small amount of lead arsenate during the projects in Nova Scotia). The French thus elected to use calcium arsenate, which they claimed "might appear to pose extremely little danger to men and animals." Nevertheless, they had still taken a number of precautions to protect the officials involved in their dusting trials, including ensuring that their pilots, their staff who loaded the dust, and their observers on the ground all wore gas masks. Furthermore, the report noted that the pilot who had flown the missions had suffered from exposure to the dust.[18]

The Canadians were given more counsel on this subject even before they had dropped one sprinkle of dust on any forest insect, but again it elicited no preventive measures. When a few of the RCAF's pilots had headed down to Tallulah, Louisiana, to train under the guidance of Bert Coad in the spring of 1927, Coad had delivered some important advice to John Wilson, the dominion's Controller of Civil Aviation, who had authority over the dusting trials in Canada. "I would like to particularly recommend," Coad instructed Wilson, "that these men be provided with the special goggles which have been developed for dusting work. The various dusts ... are hard on the eyes and absolutely require dust-proof goggles." Coad even explained to the Canadian pilots where they could purchase this specialized personal protective equipment.[19] Then upon returning from Louisiana, Flying Officer T.M. Shields declared categorically that "the pilot cockpit must be made dust proof in order to prevent dust leaking into same, all cable holes, etc., in the sides of the fuselage must be covered over with fabric. Pilots engaged on this work should always wear Meyrewitz goggles," just as Coad had insisted. The evidence presented in the preceding chapters and in what follows makes clear, however, that the goggles were not purchased and the cockpits in the planes that the Canadians used for their aerial dusting projects were anything but sealed.[20]

Even when the dusts were shown to be endangering the lives of the humans directly involved in the aerial operations, Canadian officials barely batted an eye; if ever there was convincing evidence of how humans see themselves as separate from nature, this was it. In 1928, for instance, the pilots on the project in Muskoka had complained that the dusts had sickened them and impaired their vision. Nevertheless, their superiors took no precautions to protect these men, even though they had openly discussed this grim state of affairs. A few years later, Malcolm Swaine, the dominion's senior forest entomologist, delivered advice about how to conduct aerial dusting operations to George Hopping, who was overseeing this work in BC. In doing so, Swaine nonchalantly told Hopping that "we nearly killed a pilot by filling his cockpit full of dust. In fact we have done it more than once."[21] Less than a week later, Hopping reassured Swaine that he was also acutely aware of the perils that came with this type of work. When Hopping and the mechanic had tried to remove the hopper from the plane, they had assumed that it had been empty, only to learn their mistake too late. "We already experienced filling the plane cabinet full of dust in our operation at the Wigwam project," Hopping explained. "My eyes became so inflamed I could not sleep and I am sure that the pilot and mechanic fared no better. For a time we could not see because our eyes were streaming. The stuff is worse than tear gas."[22]

These traumatic experiences eventually spurred Canadian officials to purchase breathing protection for those involved in the final aerial dusting projects, but the measures were grossly inadequate. When the dominion's Air Board conducted trials in the Ottawa area in 1931 in an effort to perfect an accurate means of delivering the desired density of toxic dusts, again officials sounded the clarion about the danger to those involved in handling the poisonous payload. Albert E. Godfrey, Wing Commander at the Ottawa Air Station, advised the Board's director in late April that, when filling the hopper, "all concerned inhale a considerable quantity of the dust, thus causing an irritation to the throat and lungs. It is therefore requested, please that four gas masks, or other suitable type respirator, be supplied for use of personnel during dusting operations."[23] In responding to this call, however, Godfrey's superiors fell well short. They secured respirators, but stressed that these were only "trainers" and not designed to stop very fine particles: "They will stop coarse smoke which might be compared to the dust complained of. Will you please warn the personnel who are using these masks that complete protection may not be given. No service respirators are available."[24]

An even more confounding situation prevailed in terms of the inability of Canadian officials to see the fatal impact the poison dusts were having on animals of all types and not simply the target insects. Swaine and his colleagues would repeatedly find dead creatures large and small, flightless and airborne, domesticated and feral, in or near the areas that they had bombed with toxins, and yet they could not or would not believe that their actions were responsible for the carnage. Again, this is all the more remarkable because Swaine and Craighead had raised the potential for this type of collateral damage in their 1924 report about the spruce budworm, and so had the French in their summary of their own project shortly thereafter.[25]

When the Canadians saw all sorts of fauna mortality in the wake of their bombing runs, however, they demonstrated an uncanny ability both to disassociate it from their dusting activities and, paradoxically, accept that such collateral damage was simply the cost of doing their business. For example, Dr. J.J. de Gryse supervised the project in Muskoka in 1928, and when he filed his report he made reference to the poison's effect on non-targeted species. "Little can be said at present concerning this," he noted, although he "noticed two dead rabbits in the dusted area shortly after the poison had been applied." de Gryse then made an observation that, while technically true, was nonetheless bewildering seeing as it was coming from one of the dominion's most esteemed senior scientists. "I cannot say with certainty," he stressed, "whether the[ir] death was due to arsenic poisoning or not."[26]

Subsequent observations and analyses were even more baffling. In early 1930, Malcolm Swaine offered a similarly perplexing analysis of the toll the dusting was taking on birds and animals when he wrote to George Hopping in BC about the observations he and his fellow officials had made in the east: "[We] watched carefully for indications of injury of this kind, [and] we have noticed actually very little of it. It is certainly to be expected that birds feeding on poisoned larvae and small animals feeding on the dusting foliage would be affected and it has been a surprise to us that we have been able to find so little evidence of anything of the sort." But he then went on to provide what was prima facie proof of just such injury. "The only animals possibly killed by the poison were the following," he recounted. "In one of our dusting operations last summer [i.e., 1929], through a misunderstanding, the end of the last hopper load was distributed broadcast and part of it went on a strip of pasture between two pieces of woodland. Two cows feeding on the pasture subsequently gave up the ghost, and the question has not yet been definitely settled, so far as I know, as to the cause of their unfortunate decease. However the owner claims that they were killed by our poison; but it was an event quite out of the ordinary and probably would not occur again in a regular dusting operation."[27]

Swaine continued to provide evidence that was at best ambivalent and at worst damning. He explained that he had never noticed any ill effects to large wild animals, like deer or moose, but postulated that "maybe there were none feeding on the dusted area." He continued in the same vein, specifically that "on two of our operations, there were noticed a small number of dead rabbits, which may possibly have been killed by the dust, but the number was so small as to be negligible."

Likewise, he asserted, "on one of our operations during the past few years, a very small number of dead birds were noticed, the figures were small, I think, two or three; but again so small as to be negligible. Unfortunately, the bodies were not sent in for a stomach examination. If there had been more of them the men would have paid them that attention."[28]

In the final analysis, even if the aerial dusting was claiming a few random vertebrates, the means justified the end as far as Swaine and his colleagues were concerned. In mid-1930, in the wake of conducting his projects in eastern Canada, Swaine admitted that "we must accept the probability of a certain injurious effect on small animals and birds from airplane dusting operations." By the same token, however, he was adamant that there was relatively little evidence of such collateral damage, and so the matter was simply not worth investigating.[29]

Arguably the most stunning example of how the Canadians involved in these aerial dusting projects were able to disregard almost completely the potential incidental damage the toxins could cause involved drinking water, and not just any old supply of it. Muskoka's recreationists rejoiced because their local lakes and rivers contained such pure, clean water, and presumably they drank it each summer while they enjoyed their seasonal retreats. Astoundingly, they expressed not one iota of concern about contaminating it with tons of poison dust in an effort to combat their hemlock looper outbreak. In fact, the cottage owners did all they could to outdo each other in the competition to ensure that their properties were coated with the toxin. Similarly, in BC in 1929, George Hopping had raised this subject as a potential concern on the cusp of overseeing the dusting operation at the Wigwam Inn (see chapter 5). He had asked officials back in Ottawa if the calcium arsenate dust he would be using would be poisonous to humans if it fell into the stream, presumably from which the inn drew its potable water. Officials in the nation's capital assured Hopping that there was little to worry about because the toxin would be so diluted in the creek. All the same, they still encouraged him to send the Dominion Chemist water samples from the creek after the dusting just to make certain, something he apparently did not do.[30]

Later that year, this issue should have become critical when Hopping was planning to dust in Seymour Canyon, an effort whose central purpose was to "save" Vancouver's internationally acclaimed pristine water supply (see chapter 6), but again it hardly raised an eyebrow. At the time, Hopping wrote to Swaine to express his concern that this operation created the very real spectre of actually poisoning the very thing it was designed to protect. Swaine candidly replied that his western colleague had raised "an important question" about which "we have at the moment no information." Swaine promised Hopping that he would review the German literature for any references to this matter and raise it with Dr. Frank T. Shutt, the Dominion Chemist.[31] Shutt replied that it was virtually impossible for him to provide an answer to Hopping because he knew nothing about the size of the local waterways and their topography and tree cover. Nevertheless, after reflecting upon the matter, Shutt concluded that, "provided the distribution of the arsenical by the airplane was carefully controlled and uniform," an aim the Canadians had perpetually struggled to achieve, the risk was "practically negligible." Although he posited that a danger would be created if a large area of exposed rock in the Canyon were dusted, whereby a subsequent rain could wash the toxin directly into what was arguably Canada's most famous supply of drinking water, he said nothing more about this subject. And with that, it was dropped.[32]

Tellingly, when Canada's aerial war on insects enjoyed its renaissance beginning in the mid-1940s, the themes discussed above remained formative in shaping events. Many authors have chronicled how the years after the Second World War saw the Americans make huge progress in their aerial war against insects. The War had expedited the development of much better aircraft and pesticides, especially dichloro-diphenyl-trichloroethane, or DDT. It was a relatively inexpensive, easy-to-use spray and effective insecticide, and was initially hailed as a potential panacea for killing pests in a wide range of settings, whether it be fire ants in the Deep South or the gypsy moth in New England. As a result, it was used wantonly and extensively.[33] Canadian forest entomologists also embraced DDT and other chemical insecticides as potential solutions in their aerial war against the pests that lurked amidst the trees. To be sure, many of their trials focused on combatting insect infestations that were menacing commercial woodlands. For example, they carried out experiments that sprayed DDT in several provinces across Canada in the mid-1940s in an effort to determine its effectiveness against a number of forest bugs. But except for the case of New Brunswick during the 1950s and 1960s, where an industry behemoth, the International Paper Company, aggressively supported what has been described as "one of the largest and longest sustained aerial insecticide spraying programs in the world," the Canadian forest entomologists conducted surprisingly few projects at the behest of the country's forest companies.[34]

Instead, much of the work they carried out, particularly in Ontario, sought to kill bugs for recreationists who felt that the pests were detracting from their enjoyment of their chosen outdoor activities; the recreationists' drive to fine-tune the splendour that nature has bestowed upon us was – and is – alive and well in postwar Canada. The damage the insects were causing varied considerably. On some occasions, it could be the traditional problem of diminishing an area's aesthetic appeal. On others, it could be that the masses of bugs were considered gross, or that the pesky, biting insects were simply a nuisance. Regardless, the same remedy was sought in each instance, whereby the planes doused the specific recreational retreats in poisons, most often DDT. It was the toxin of choice for forest entomologists in their first trial, which they conducted against the spruce budworm in Algonquin Park, the crown jewel of Ontario's legendary provincial park system, in 1944. Over the next few decades, they oversaw projects that showered forest pests with toxins in Muskoka, Haliburton, and Georgian Bay (to kill forest tent caterpillars), and in other fabled provincial parks in Ontario such as Quetico and Rondeau (for spruce budworm and mosquitoes). The forest entomologists also conducted another treatment in Stanley Park in BC in 1959 when the western hemlock looper returned with a vengeance.[35]

Herein lies the ultimate lesson that we can draw from recounting the history of Canada's pioneering aerial campaign against forest insects between 1927 and 1930. For at least the last half century, environmental preservationists and the forest industry have generally been at loggerheads when it comes to managing our woods. One group appreciates the trees only as long as they are standing because of their perceived aesthetic and psychological value. The other prizes the woods only as long as it can cut them in the future and convert them into commercial products. Given these agendas, conflict has understandably defined the relationship between the two sides. And yet, both of them were actuated by the same animus – self-interest – to kill bugs that were damaging their trees. Both saw dumping poisons on the pests as the best means to realizing their ends and getting what they wanted from the woods.

When it came to the case each side presented to explain why taking such action made sense, however, these two groups could not have been more different. The forest industry's officials – indeed, everyone involved in the projects we have examined that dealt with commercial woods – were remarkably forthright when they presented their case for shaping environmental policy. Their arguments focused on comparing the tangible timber values that they could realize from harvesting a stand against the cost of treating it with pesticides. If the former were greater than the latter then the price tag was worth it. This was logical reasoning at its finest. Not so for the trees that were worshipped for their aesthetic qualities, however. There was a general consensus that such straightforward arithmetic was completely inappropriate when determining if the government should dust these trees because it was agreed – practically universally – that forests that were worshipped for their aesthetic qualities were of incalculable value. Dusting them with toxins was always worth the cost. Moreover, those who lobbied for such measures infused their messages with pronounced moralistic tones, whereby they argued that "saving" a stand of beautiful trees was simply the "right" thing to do. By tugging at heart strings, these entreaties capitalized on the juggernaut power of emotional appeals to wash away any potential divisions of class and thereby exercise enormous influence in shaping environmental policy.

That is the rub of the story of Canada's aerial dusting campaign against forest pests during the interwar years. It poignantly illustrates the grave dangers inherent in allowing emotion and not logic to determine how we manage nature. In a variety of settings across Canada, environmental preservationists who espoused a profound love of the woods fervently argued that it behooved public officials to bomb bug-infested woods with deadly dusts in order to save the trees. The preceding pages have already described many of the myriad absurdities that marked their actions and words, the most noteworthy of which was the claim that the only way to save nature was to kill it. But the strength of their lobby also drove government officials to take extreme and reckless actions to address their concerns, barely thinking twice about doing so. This included forcing their pilots to fly missions that were doomed to fail and which they knew would probably also kill these aviators, and dumping deadly poisons into drinking water supplies, one of which was arguably the purest in Canada. Furthermore, elected officials had little choice but to respond to the calls to dust certain sites for aesthetic reasons because the demands came from a segment of society that carried such an inordinate amount of political weight. Rare is the principled politician who can resist such impassioned pleas for help, especially when they emanate from influential power brokers.

This dynamic is worrisome because conditions are conducive to it continuing to be a major force shaping natural resource management policy today. As Canadian society becomes increasingly urbanized and detached from the country's lakes and trees and birds and animals, the manner in which the nation's collective psyche conceives of nature is based less and less upon a first-hand, logical understanding of it. Instead, its foundation increasingly consists of a romanticized notion of how the flora and fauna around us ought to look and behave. These circumstances are more likely to elicit emotional and extreme reactions to the environmental issues we confront. As Tina Loo puts it, these conditions drive us to "cultivate a way of seeing and being that precludes forging a truly sustainable relationship with the environment."[36] We would thus do well to remind ourselves of the lessons provided by the story of Canada's aerial campaign against forest pests during the interwar years, principally the need to let our heads and not our hearts influence our policies and actions as we endeavour to tackle the pressing environmental problems that lie ahead in our contemporary world.

Notes

Introduction

1 Archives of Ontario (AO), RG1–256, Box 1, Hemlock Looper (1), 4 October 1927, H. Balm to Forestry Department.
2 AO, RG1–256, Box 1, Hemlock Looper (1), 30 August 1927, E.J. Zavitz to G. Freemantle.
3 Swaine, *The Spruce Budworm*; Swaine, *The Hemlock Looper*.
4 I am grateful to the second reviewer who read my manuscript and suggested that I incorporate into it a discussion of lobbying activity. This person provided immensely helpful advice about sources to consult, such as Deschouwer and Jans, *Politics Beyond the State*.
5 This argument resonates with the views William Cronon presents in his iconic article, "The Trouble with Wilderness."
6 Andrews, *Timber*, 13.
7 "Aldo Leopold," The Aldo Foundation, accessed 25 May 2018, https://www.aldoleopold.org /about/aldo-leopold; Leopold, *A Sand County Almanac*, 101.
8 Cronon, "The Trouble with Wilderness."
9 For example, see Nash, *Wilderness and the American Mind*; Hays, *Conservation and the Gospel of Efficiency*; Hays, *Beauty, Health and Permanence*.
10 Leopold, *A Sand County Almanac*, 107–8.
11 Schneider, *Hybrid Nature*; White, *The Organic Machine*; see also Nash, "The Changing Experience of Nature"; Gandy, *Concrete and Clay*.
12 See Spirn, "Constructing Nature"; Favareau, "Understanding Natural Constructivism."
13 Cronon, "The Trouble with Wilderness," 16–17.
14 Writers in Canada, just as in the United States, have been chronicling the non-human world and reflecting upon our relationship with it for centuries. Nearly 200 years ago, for instance, Catharine Parr Traill published *The Backwoods of Canada*, her account of life in what was then Upper Canada in British North America. About a century later, Arthur R.M. Lower published three books that describe in great detail Canada's forests, their exploitation, and the society that the timber trade created. In this book, however, environmental history refers to the discrete sub-field that developed in Canada around the turn of the twenty-first century.

15 Jasen, *Wild Things*; MacEachern, *Natural Selections*; Campbell, *Shaped by the West Wind*; Loo, *States of Nature*; Kheraj, *Inventing Stanley Park*; Kheraj, "Restoring Nature"; Parr, *Sensing Changes*. See also Macfarlane, "'A Completely Man-Made and Artificial Cataract'"; Cruikshank and Bouchier, *The People and the Bay*.

16 Zeller, *Inventing Canada*; Zeller, *Land of Promise, Promised Land*; Girard, *L'écologisme retrouvé*; Whorton, *Before Silent Spring*; Carson, *Silent Spring*; Dunlap, *DDT*; Russell, "Speaking of Annihilation"; Russell, *War and Nature*. See also Buhs, *The Fire Ant Wars*.

17 Castonguay, *Protection des cultures*; Castonguay, "Naturalizing Federalism"; Castonguay, "The Emergence of Research Specialities."

18 Cook, "'Spray, Spray, Spray!'" 8.

19 Evans, *The War on Weeds in the Prairie West*; Riegert, *From Arsenic to DDT*; Rajala, "The Vernon Laboratory"; Gibson, "The Canadian Entomological Service"; Turner, "A History of Investigations"; Prebble, *Aerial Control of Forest Insects in Canada*; West, *The Firebirds*, 147–50. A few works deal with more recent issues involving forest pests in Canada; for example, see McLaughlin, "Green Shoots"; Sandberg and Clancy, "Politics, Science and the Spruce Budworm."

20 Creighton, *The Commercial Empire of the St. Lawrence*; Bliss, *Northern Enterprise*.

21 Forster, *A Conjunction of Interests*; Traves, *The State and Enterprise*; Cruikshank, *Close Ties*; Lower, *Settlement and the Forest Frontier in Eastern Canada*.

22 Macdonald, *Business and Environmental Politics in Canada*; McLaughlin, "Green Shoots"; Sandberg and Clancy, "Politics, Science and the Spruce Budworm."

23 One such study even contends that, "before the Second World War, there was little support among forest scientists for chemical controls": Sandberg and Clancy, "Politics, Science and the Spruce Budworm," 173.

1 "Airplane dusting offers the only present hope"

1 Pyne, *Fire in America*, 71; Krech, "Fire."

2 Entomologists in other fields drew the same conclusions at this time. For example, those involved in tackling pest problems on Western Canada's grasslands deduced that poor land-use practices were the root cause of the local infestations, and that the only abiding solution was to promote a healthier approach to managing these environments. Thistle, "Accommodating Cattle," 70–2.

3 Castonguay, *Protection des cultures*, chap. 1; Riegert, *From Arsenic to DDT*, chaps. 1–4; Whorton, *Before Silent Spring*, 3–7; Russell, *War and Nature*, 2–7.

4 Zeller, *Inventing Canada*, 271.

5 Cook, "'Spray, Spray, Spray!'"

6 Castonguay, *Protection des cultures*, chap. 1; Riegert, "James Fletcher"; Riegert, *From Arsenic to DDT*, chaps. 1–4; Gibson, "The Canadian Entomological Service."

7 "Arthur Gibson," Find a Grave, accessed 4 May 2018, https://www.findagrave.com/memorial /147596045/arthur-gibson.

8 de Vecchi, "Science and Government in Nineteenth-Century Canada," describes how, in the early twentieth century, the dominion bureaucracy provided an intellectual climate that was conducive to abandoning science based on data collection and adopting a more interventionist, experimental approach. Evans, *The War on Weeds in the Prairie West*, chronicles the parallel battle against undesirable plants that scientists fought at this time.

9 Gibson, "The Canadian Entomological Service," 1433–4; Castonguay, *Protection des cultures*, chap. 1; Riegert, *From Arsenic to DDT*, chap. 4.

10 Castonguay, "Naturalizing Federalism," 20–5: the author describes how the dominion Department of Agriculture was able to concentrate the technical entomological manpower within its ranks and thereby justify its encroachment upon a field of jurisdiction traditionally occupied by provincial governments.

11 Riegert, "Charles Hewitt Gibson"; Castonguay, *Protection des cultures*, chap. 1 and pp. 62–5; Gibson, "The Canadian Entomological Service," 1440; Rajala, "The Vernon Laboratory," 177.

12 Castonguay, "Naturalizing Federalism," 16.

13 Castonguay, *Protection des cultures*, 60–1; Swaine and Craighead, *Studies on the Spruce Budworm*, 4–8; Hewitt, *Report of the Dominion Entomologist*, 235, 245; Riegert, *From Arsenic to DDT*, chap. 6.

14 White pine suffered enormous damage from several sources in the early twentieth century. A native pest, the weevil, would be "discovered" at this time, as would blister rust, which was a foreign pathogen that was inadvertently imported from Europe.

15 Swaine, *Forest Entomology*, 1; Kuhlberg, *In the Power of the Government*, 3–82.

16 Swaine and Craighead, *Studies on the Spruce Budworm*, 16–22; Swaine, *The Spruce Budworm*; Natural Resources Canada, "Spruce Budworm (Factsheet)," Government of Canada, accessed 18 November 2016, http://www.nrcan.gc.ca/forests/fire-insects-disturbances/top-insects/13403; Prebble, *Aerial Control of Forest Insects in Canada*, 81.

17 Gibson, "The Canadian Entomological Service," 1434–5. Fletcher first described the damage allegedly caused by the budworm in New Brunswick in 1885, but he suspected that the injury to the trees at that time was primarily caused by spruce bark beetles.

18 "James Malcolm Swaine," 460; Gibson, "The Canadian Entomological Service," 1436; Email correspondence about James M. Swaine with Hilary Dorsch Wong, Reference Coordinator, Cornell University Archives (CUA) – Rare Books and Manuscripts, 14 June 2016; Castonguay, *Protection des cultures*, 126.

19 Castonguay, *Protection des cultures*, 52, 126–8; Gibson, "The Canadian Entomological Service," 1437; Swaine and Craighead, *Studies on the Spruce Budworm*, 7–9; Rajala, "The Vernon Laboratory," 177–8.

20 MacEachern, *The Miramichi Fire*.

21 Kuhlberg, *One Hundred Rings and Counting*, chaps. 1–2; Pyne, *Awful Splendour*, 87, 127–31.

22 Swaine, *Canadian Bark-Beetles*, 3. Hewitt and Swaine repeatedly delivered this message whenever they could, including at a series of public addresses. See also Hewitt, *The Spruce Budworm and Larch Sawfly*; Hewitt, *The Control of Insect Pests in Canada*; Canada Commission of Conservation, *First Annual Report 1910*, 142–52. American forest entomologists were faced with the same challenging situation during the interwar period, and they complained about it to some of their Canadian allies: Nova Scotia Archives (NSA), RG20, 720, Pe-x, 16 March 1926, H.B. Pierson to F.J.D. Barnjum and 12 April 1926, H.B. Pierson to O. Schierbeck.

23 Canada Department of Agriculture, *Annual Report of the Entomologist and Botanist for the Calendar Year 1886*, passim and all the reports for the subsequent two decades; Cook, "'Spray, Spray, Spray!'" passim; Riegert, *From Arsenic to DDT*, chaps. 4–7; Castonguay, *Protection des cultures*, 32.

24 Castonguay, *Protection des cultures*, chaps. 1–2; Whorton, *Before Silent Spring*, chap. 1, describes how the same situation prevailed in the United States.

25 Turner, "A History of Investigations," 1–6; Hopping, "The Western Hemlock Looper," 4.

26 Jansen, "Chemical-Warfare Techniques for Insect Control"; Douglas, *The Creation of the National Air Force*, chap. 1.

27 To be sure, timber cruisers would still walk the forests to collect the detailed data that observers could not produce from the cockpits of the planes. The true value of the aircraft was their ability

to provide foresters with a general overview of the state of large tracts of their woodlands very quickly.

28 Douglas, *The Creation of the National Air Force*, chaps. 1–2; Greenhous and Halliday, *Canada's Air Forces*, 20–4.

29 Library and Archives Canada (LAC), MG30-E243, Finding Aid "Biography"; LAC, MG30-E322, Finding Aid "Biography"; LAC, MG30-E322, Scrapbook – General, newspaper clippings from 1920s; LAC, RG24-E-1-a, 4898, 1008-2-2, vols. 1–2, all documents; Douglas, *The Creation of the National Air Force*, 40–1.

30 Wilson, "The Use of Aircraft in Forestry and Logging"; Wilson, "Through Canadian Wilds," 16–25; Douglas, *The Creation of the National Air Force*, 63–6.

31 Gibson, "The Canadian Entomological Service," 1469. Hearle would continue this work over the course of 1920–1 and become an expert in the field of mosquito control. Riegert, *From Arsenic to DDT*, chap. 8.

32 Gibson, *A Report of the Dominion Entomologist*, 19.

33 LAC, RG17, 3047, 40–6(1), PC2202; Gibson, *Report of the Dominion Entomologist*, 18–19; Castonguay, *Protection des cultures*, 136; United States Department of Agriculture, *Trees*, 471.

34 LAC, RG24-E-1-a, 4912, DND – Operations for Department of Agriculture, Vol. 1, 27 September 1921, J.M. Swaine to J.A. Glen; 30 January 1923, H.T. Gussow, Memorandum; 2 February 1923, J.H. Grisdale to G.J. Desbarats; 7 January 1924, H.T. Gussow to E.L. Godfrey; and 24 January 1924, A.T. Davidson to E.L. Godfrey. See also Canada Department of Agriculture, Entomological Branch, *The Canadian Insect Pest Review* 2, no. 3 (June 1924): 26; 3, no. 1 (April 1925): 7; 3, no. 4 (August 1925): 32; and 3, no. 5 (September 1925): 40.

35 For example, see Canada Department of Agriculture, Entomological Branch, *The Canadian Insect Pest Review* 1, no. 1 (April 1923).

36 LAC, RG24-E-1-a, 4912, DND – Operations for Department of Agriculture, Vol. 1, 11 December 1922, A. Gibson to J.H. Grisdale.

37 LAC, RG24-E-1-a, 4912, DND – Operations for Department of Agriculture, Vol. 1, 8 March 1923, G.O Johnson, Flight Lieutenant, Air Board – Memorandum of Discussions with Dr. Swaine re: Operations for the Department of Agriculture.

38 J.M. Swaine, "Progress in Forest Insect Control in Canada," *Forestry Chronicle* 4, no. 1 (1928): 35–40.

39 Department of the Interior, Forestry Branch, "Report of the Superintendent of Forestry," 16 July 1908 (in *Sessional Paper XXV*, Department of the Interior, in *Sessional Papers, Vol. XIV*, First Session of the Eleventh Parliament of the Dominion of Canada, Session 1909), 6, from which the first citation is taken; LAC, RG17, 3131, 69(1) – Militia and National Defence, 5 December 1924, J.M Swaine to J.H. Grisdale, enclosing "The Present Status of Airplane Dusting Over Forest Land," from which the second citation is taken.

40 LAC, RG17, 3037, Destructive Insect and Pest Act (2), 23 May 1923, W.R. Motherwell to C.M. Hamilton; see also 21 May 1923, C.M. Hamilton to W.R. Motherwell.

41 Swaine and Craighead, *Studies on the Spruce Budworm*, 88.

42 LAC, RG24-E-1-a, 4912, DND – Operations for Department of Agriculture, Vol. 1, 8 March 1923, G.O. Johnson, Memorandum; 16 August 1923, J.M. Swaine to M. Stedman; and 26 November 1923, J.M. Swaine to J.A. Wilson.

43 Neillie and Houser, "Fighting Insects with Airplanes," 338. Downs and Lemmer, "Origins of Aerial Crop Dusting," 123–5, describe how the Americans conducted a few early dusting trials over "forested areas," but the treed tracts that they treated – such as the one in Ohio – were truly groves or rows of trees in urban or rural settings and not natural forests.

44 The cotton leaf worm was also a problem, but the boll weevil was the primary concern.

45 Downs and Lemmer, "Origins of Aerial Crop Dusting," 126–7; Post, "Boll Weevil Control by Airplane"; Coad et al., *Dusting Cotton from Airplanes*; Lecky and Murphy, *History of Tallulah Laboratory*.

46 LAC, RG17, 3131, 69(1) – Militia and National Defence, 5 December 1924, J.M Swaine to J.H. Grisdale, enclosing "The Present Status of Airplane Dusting Over Forest Land."

47 Lecky and Murphy, *History of Tallulah Laboratory*; Neillie and Houser, "Fighting Insects with Airplanes," 333; Post, "Boll Weevil Control by Airplane," passim; Coad et al., *Dusting Cotton from Airplanes*; "Huff-Daland-Duster," Delta Flight Museum, accessed 9 December 2011, https://www.deltamuseum.org/exhibits/delta-history/aircraft-by-type/crop-duster/Huff-Daland-Duster; Hopping, "The Western Hemlock," 44–5; LAC, RG24-E-1-a, 4912, DND – Operations for Department of Agriculture, Vol. 3, 10 February 1928, Secretary to British Air Attaché in Berlin to E.W. Stedman, enclosing "Fighting of Insect Pests"; LAC, RG12, 3037, Destructive Insect and Pest Act (3), 22 April 1925, J.M. Swaine to W.R. Motherwell.

48 The resolution was reprinted in *Mail and Empire*, 10 July 1925.

49 Downs and Lemmer, "Origins of Aerial Crop Dusting," 126; LAC, RG17, 3131, 69(1) – Militia and National Defence, 5 December 1924, J.M Swaine to J.H. Grisdale, from which the citation is taken.

50 LAC, RG17, 3131, 69(1) – Militia and National Defence, 10 December 1924, J.M. Swaine to J.A. Wilson.

51 LAC, RG17, 3131, 69(1) – Militia and National Defence, 29 January 1925, J.A. Wilson to J.M. Swaine. See also Douglas, *The Creation of the National Air Force*, 77.

52 LAC, RG17, 3131, 69(1) – Militia and National Defence, 16 February 1925, J.M. Swaine to J.A. Wilson.

53 LAC, RG17, 3131, 69(1) – Militia and National Defence, 19 February 1925, J.H. Grisdale to Director of Air Service, which is a verbatim copy of LAC, RG17, 3131, 69(1) – Militia and National Defence, 16 February 1925, J.M. Swaine to J.H. Grisdale; Grisdale apparently permitted Swaine to compose the letter he sent to the RCAF. See also LAC, RG17, 3131, 69(1) – Militia and National Defence, 21 April 1925, A. Gibson to J.H. Grisdale, Memorandum of Flying Operations for the Entomology Branch.

54 Post, "Boll Weevil Control by Airplane," 14.

55 Coad et al., *Dusting Cotton from Airplanes*, 35. Coad remained convinced that aerial dusting over forests was unsafe even after the Canadians had begun doing so. He wrote in the late 1920s that dusting aircraft had their limitations, and that they could only be used "in areas where conditions are suitable. Reasonably level land is absolutely essential and fairly treeless areas are highly desirable": United States Department of Agriculture, *Yearbook of Agriculture 1928*, 119.

56 Neillie and Houser, "Fighting Insects with Airplanes," 332; Post, "Boll Weevil Control by Airplane," 16.

57 LAC, RG17, 3131, 69(1) – Militia and National Defence, 18 March 1925, G.J. Desbarats to J.H. Grisdale. See also LAC, RG24, 5064, 1021-1-26, 4 March 1925, B.D. Hobbs to Director; LAC, RG24-E-1-a, 4912, DND – Operations for Department of Agriculture, Vol. 1, 17 March 1925, J.S. Scott, Memorandum.

58 LAC, RG17, 3131, 69(1) – Militia and National Defence, 30 March 1925, G.J. Desbarats to J.H. Grisdale, which is enclosed in 1 April 1925, J.H. Grisdale to A. Gibson. Although the Air Board did not conduct Swaine's aerial dusting project that season, it carried out a slew of other projects for the Entomological Branch. See LAC, RG17, 3131, 69(1) – Militia and National Defence, List of requests for flying operations for 1925 received from other government departments, complete up to 25 February 1925.

59 LAC, RG24-E-1-a, 4912, DND – Operations for Department of Agriculture, Vol. 1, 26 May
    1925, J. Lawler to J.A. Wilson, enclosing "Dust Peach Orchard from Special Plane," *Globe*, May
    25, 1925; LAC, RG24-E-1-a, 4912, DND – Operations for Department of Agriculture, Vol. 1, 3
    June 1925, A.L. Morse to G.E. Wait.

60 LAC, RG17, 3037, Destructive Insect and Pest Act, Vol. 3, 22 April 1925, J.M. Swaine to W.R.
    Motherwell.

61 Edmund Russell is the best known among several authors who have chronicled the remarkable
    degree to which the strategies, technologies, and metaphors of war shaped the approaches to
    and understandings of insect control in North America during the twentieth century. Russell,
    "Speaking of Annihilation"; Russell, *War and Nature*; see also Buhs, *The Fire Ant Wars*.

62 Dominion of Canada, *Official Report of Debates*, House of Commons, Fourth Session –
    Fourteenth Parliament, 15-16 George V, 1925, Volume IV, 1925, 3263.

63 "The Greatest War of All," *Maclean's*, September 1, 1925.

64 LAC, RG17, 3037, Destructive Insect and Pest Act – General (3), 2 September 1925, W.R.
    Motherwell to J.V. MacKenzie.

65 LAC, RG24-E-1-a, 4912, DND – Operations for Department of Agriculture, Vol. 1, 18
    November 1925, J.H. Grisdale to G.J. Desbarats, enclosing Memorandum re: Flying Operations
    for 1926.

66 LAC, RG17, 3131, 69(1) – Militia and National Defence, 14 November 1925, A. Gibson to J.H.
    Grisdale, as well as their correspondence from late 1925 and early 1926; LAC, RG24-E-1-a, 4912,
    DND – Operations for Department of Agriculture, Vol. 1, 22 December 1925, J.H. Grisdale to
    G.J. Desbarats.

67 J.M. Swaine, "The Present Status of the Spruce Budworm Outbreaks in Eastern Canada," *Forestry
    Chronicle* 2, no. 3 (1926): 34–6.

68 Ibid.

69 Canada Department of Agriculture, Entomological Branch, *The Canadian Insect Pest Review* 4,
    no. 3 (25 July 1926): 25; 4, no. 4 (31 August 1926): 36; and 4, no. 5 (30 October 1926): 44. See
    also LAC, RG17, 3131, 69(1) – Militia and National Defence, 4 December 1926, A. Gibson to
    J.H. Grisdale; 10 December 1926, J.H. Grisdale to G.J. Desbarats; and 18 December 1926, W.W.
    Cory to J.H. Grisdale.

70 LAC, RG17, 3131, 69(1) – Militia and National Defence, 25 November 1926, E.S. Archibald
    to Acting Deputy Minister of Agriculture, enclosing 24 November 1926, H. Gussow,
    Memorandum. See also LAC, RG17, 3131, 69(1) – Militia and National Defence,
    11 November 1926, G.J. Desbarats to J.H. Grisdale; Douglas, *The Creation of the National
    Air Force*, 116.

71 LAC, RG24-E-1-a, 4912, DND – Operations for Department of Agriculture, Vol. 2, 28 December
    1926, J.S. Scott to Chief of Staff. See also LAC, RG24-E-1a, 5070, 1021–1-77, 25 October 1926,
    Memorandum for the Ministers of Interior and National Defence; 22 October 1926, Draft
    Order-in-Council; and 6 December 1926, PC 2023; LAC, RG24-E-1-a, 4912, DND – Operations
    for Department of Agriculture, Vol. 2, 15 December 1926, Chief of Staff, DND to Director of
    RCAF; and 16 December 1926, J.S. Scott to T.G. Hetherington.

72 LAC, RG17, 3131, 69(1) – Militia and National Defence, 4 December 1926, A. Gibson to J.H.
    Grisdale. See also LAC, RG24-E-1-a, 4912, DND – Operations for Department of Agriculture,
    Vol. 2, 12 November 1926, G.J. Desbarats to J.H. Grisdale.

73 LAC, RG17, 3131, 69(1) – Militia and National Defence, 10 December 1926, J.H. Grisdale to G.J.
    Desbarats and W.W. Cory.

74 LAC, RG17, 3131, 69(1) – Militia and National Defence, 18 December 1926, W.W. Cory to J.H.
    Grisdale.

75 LAC, RG39, 341, 47278, 11 November 1926, Inspector of Water and Forests, Mr. Noel [translation]; Jansen, "Chemical Warfare Techniques for Insect Control."

76 LAC, RG39, 341, 47278, 11 November 1926, Inspector of Water and Forests.

77 LAC, RG24-E-1-a, 4912, DND – Operations for Department of Agriculture, Vol. 1, 27 April 1926, J.M. Swaine to J.A. Wilson.

78 Apparently the Germans, specifically firms such as Junkers Flugzeugwerk A-G, were at the forefront in developing this technology and had done so in a way that inextricably linked their hoppers to their planes, thereby making it exponentially more expensive to conduct aerial dusting projects. See LAC, RG24-E-1-a, 4912, DND – Operations for Department of Agriculture, Vol. 1, 15 April 1926, J.A. Wilson to J.M. Swaine; LAC, RG24-E-1-a, 4913, DND – Operations for Department of Agriculture, Vol. 4, 8 April 1929, W. Knight (Vice-president, Junkers Corporation of American) to J.M. Swaine.

79 The Americans may have proven so receptive to the Canadians' request because the latter had gone through the official channels, namely their British representatives in Washington. A few years earlier, when the Canadians had directly asked the Americans for an estimate of the cost of using a blimp for aerial dusting, their inquiry went unanswered. LAC, RG24-E-1-a, 4912, DND – Operations for Department of Agriculture, Vol. 1, 29 April 1925, M.G. Christie to G. Wait; 4 May 1925, G. Wait to RCAF Liaison Officer, London, England; 9 June 1925, N.E. Margetts to Acting Director, RCAF; and 6 July 1925, British Air Ministry to Secretary, [Canadian] Department of National Defence.

80 LAC, RG17, 3131, 69(1) – Militia and National Defence, 20 December 1926, G.J. Desbarats to J.H. Grisdale; LAC, RG24-E-1-a, 4912, DND – Operations for Department of Agriculture, Vol. 2, 4 January 1927, T.G. Hetherington to Director, J.S. Scott, enclosing Coad et al., *Dusting Cotton from Airplanes*, and Neillie and Houser, "Fighting Insects with Airplanes."

2 "One of the first aerial applications of an insecticide in forestry"

1 Schmitt et al., *Managing the Spruce Budworm*, 116.

2 Chroniclers of Nova's Scotia's forest history have said almost nothing about this aerial dusting project in their province. For example, Hawboldt, *The Spruce Budworm*, traces the spruce budworm infestations with which the province dealt during the mid-1920s without noting the effort described in this chapter to combat them; likewise, Sandberg and Clancy, "Politics, Science and the Spruce Budworm." One source that mentions it is Johnson, *Forests of Nova Scotia*, 166.

3 Hosie, *Native Trees of Canada*, 22–3; Fernow, *Forest Conditions of Nova Scotia*; *Report of the Royal Commission on Pulpwood*, chap. 2. The following paragraph is also based upon these sources.

4 Examining the timber maps from the Commission of Conservation's investigation of Nova Scotia's timber resources in 1912 provides a remarkable illustration of the conspicuousness of the Big Lease. Its uniform markings and large size are a glaring anomaly amidst a veritable ocean of heterogeneous and fragmented timber types and age classes, burns, and wetlands: Fernow, *Forest Conditions of Nova Scotia*.

5 Sandberg, "Forest Policy in Nova Scotia."

6 *Report of the Royal Commission on Pulpwood*, 26.

7 "Otto Schierbeck," *Forestry Chronicle* 17, no. 3 (1941): 134.

8 Roach and Judd, "A Man for All Seasons"; Parenteau and Sandberg, "Conservation and the Gospel of Economic Nationalism."

9 Reviewing the correspondence between Barnjum and Schierbeck leaves no doubt that the former saw the latter as his means of realizing his industrial designs in Nova Scotia's forest industry.

10 It is unclear if the dominion entomologists felt slighted by Schierbeck's attempt to steal their thunder by publishing his essay, which focused on the budworm, while they were still compiling the data for their groundbreaking report about this insect (Swaine and Craighead, *Studies on the Spruce Budworm*), which they published a short time later. Tellingly, Schierbeck openly admitted in his treatise that he had worked with Craighead on studying the budworm, but neither Craighead nor Swaine recognized Schierbeck's contribution to their effort when they published their document.

11 Schierbeck, *Treatise on the Spruce Bud Worm*, passim with the citation taken from 28–9.

12 Ibid., 31.

13 "$5,000,000 Paper Mill for Queens," *Halifax Herald*, June 9, 1925; Roach and Judd, "A Man for All Seasons."

14 "Budworm Ravaging Forests in Cape Breton District," *Globe*, August 28, 1925.

15 Canada Department of Agriculture, Entomological Branch, *The Canadian Insect Pest Review* 2, no. 7 (October 1924): 76; LAC, RG17, 3047, 40–6(1), Report on Spruce Budworm Conditions in Cape Breton Island, December 1925, M.B. Dunn; NSA, RG20, 720, Y, 25 September 1925, J.M. Swaine to O. Schierbeck.

16 "Budworm Scourge Menaces Entire Coniferous Forests," *Globe*, September 18, 1925; "Forests of Cape Breton Are Being Ruined by Plague," *Halifax Herald*, September [19], 1925. Ironically, in light of future events, Schierbeck carried out a supplementary investigation of a budworm outbreak in Guysborough County and was "confident that there was no outbreak" there: NSA, RG20, 718, Attorney-General – Insects and Fungi, 29 September 1925, O. Schierbeck to J.C. Douglas; NSA, RG20, 746, Budworm, 29 September 1925, O. Schierbeck to J.C. Douglas, from which the citation is taken.

17 "Forests of Cape Breton in Greatest Jeopardy if Budworm Not Exterminated," *Sydney Record*, September 19, 1925. The *Halifax Herald* repeated this extreme rhetoric a short while later: see *Halifax Herald*, September 28, 1925.

18 "Budworm Is Doing Much Damage to Pulpwood Forests of Cape Breton," *Canada Lumberman*, October 1, 1925.

19 NSA, RG20, 720, Y, 25 September 1925, J.M. Swaine to O. Schierbeck. Unfortunately, relations between Swaine and Schierbeck were not helped when a miscommunication that occurred between the two sent a dominion entomologist to the wrong location, thereby delaying the survey of the budworm situation in Nova Scotia. Schierbeck interpreted this as yet another snub of Nova Scotia's interests, whereas the evidence makes it clear that it was an honest mistake: NSA, RG20, 718, Attorney-General – Insects and Fungi, 16 November 1925, J.M. Swaine to O. Schierbeck; NSA, RG20, 746, Budworm, 25 November 1925, J.M. Swaine to J.C. Douglas; NSA, RG20, 719, Du-Dz, 28 November and 10 December 1925, O. Schierbeck to M.B. Dunn.

20 It is likely that Schierbeck recommended this control method because of his experience in dealing with beech louse, which was now an invasive species in Nova Scotia after having migrated from Europe. It was a very slow-moving insect because of its short legs, and outbreaks of it could thus be contained by cutting infested trees and those around them. In contrast, spruce budworm moths could fly significant distances, thereby rendering such control methods useless: NSA, RG20, 746, Budworm, 18 November 1925, O. Schierbeck to J.A. Walker.

21 NSA, RG20, 718, Attorney-General – Insects and Fungi, 5 November 1925, O. Schierbeck to J.C. Douglas; NSA, RG20, 746, Memoranda – Miscellaneous, ca. 5 November 1925, Memorandum regarding Meeting of the Attorney-General, Otto Schierbeck and the Timber Operators from the Area of the Proposed Control Cutting. It is unclear if the tax was to be a one-time or annual toll. In any event, the idea of such an unexpected levy was anathema to the province's timber owners.

22 NSA, RG20, 746, Memoranda – Miscellaneous, ca. 5 November 1925, Memorandum regarding Meeting of the Attorney-General, Otto Schierbeck and the Timber Operators from the Area of the Proposed Control Cutting; NSA, RG20, 720, Me-Mh, 23 October 1925, O. Schierbeck to Oxford Paper Company and 26 October 1925, H. Chisholm to O. Schierbeck; NSA, RG20, 720, Sn-So, O. Schierbeck to Sonora Lumber Company and other large timber holders on Cape Breton Island; NSA, RG20, 746, Budworm, 23 and 30 October 1925, O. Schierbeck to J.C. Douglas.

23 Maine Historical Society, Collection 1882, Fonds Description; *Lewiston Daily Sun*, September 3, 1937; Leane, *The Oxford Story*, passim.

24 I am thankful to Senator Susan M. Collins, Justin Davis (her Staff Assistant), and Jerry W. Mansfield (Information Research Specialist – Knowledge Services Group at the Congressional Research Service) for their help in September 2012 in trying to unravel how Craighead was granted this special leave. Although the search turned up empty, their assistance was immensely appreciated.

25 NSA, RG20, 720, Me-Mh, 13 and 23 October 1925, O. Schierbeck to H.J. Hugill and Oxford Pulp and Paper Company; NSA, RG20, 746, Memoranda – Miscellaneous, ca. 5 November 1925, Memorandum regarding Meeting of the Attorney-General, Otto Schierbeck and the Timber Operators from the Area of the Proposed Control Cutting; NSA, RG20, 746, Budworm, 3 November 1925, O. Schierbeck to F.C. Craighead.

26 NSA, RG20, 718, Attorney-General – Insects and Fungi, 11 November 1925, F.C. Craighead, "Report on Examination of the Budworm in Cape Breton." Schierbeck had even been the one to personally invite Craighead to Nova Scotia: see NSA, RG20, 746, Budworm, 3 November 1925, O. Schierbeck to F.C. Craighead. Craighead sent a copy of his report to Swaine a short while later: NSA, RG20, 746, Cr-Cl, 16 November 1925, F.C. Craighead to J.M. Swaine and copy to O. Schierbeck.

27 NSA, RG20, 746, Budworm, 25 November 1925, J.M. Swaine to J.C. Douglas.

28 NSA, RG20, 718, Attorney-General – Insects and Fungi, 12 November 1925, J.M. Swaine to O. Schierbeck. Over the previous fortnight or so, John Douglas, Nova Scotia's de facto Minister of Lands and Forests, repeatedly asked Swaine to send immediately an entomologist to Nova Scotia to investigate the budworm situation. Swaine's absence delayed this from happening, although in mid-November he promised Douglas to send someone as soon as conditions permitted: NSA, RG20, 746, Budworm, 3 November 1925, A. Gibson to O. Schierbeck; 4 November 1925, C. Stewart to J.C. Douglas; and 11 and 12 November 1925, J.M. Swaine to J.C. Douglas.

29 United States National Archives and Records Administration – Kansas City, ARC Identifier 2092708, RG7, General Correspondence, 1925–1934, Box 273, Swaine J.M., 16, 23 and 30 November 1925, F.C. Craighead to J.M. Swaine.

30 NSA, RG20, 718, Attorney-General – Insects and Fungi, 8 December 1925, O. Schierbeck to J.C. Douglas.

31 Ibid.

32 Significantly, Schierbeck took a shot at S.A. Graham, an American entomologist, who had argued that slash disposal was not an effective practice in combatting beetle infestations. This was significant because Frank Craighead fervently agreed with Graham, and no doubt this further soured relations between him and Schierbeck: see S.A. Graham, "Some Entomological Aspects of the Slash Disposal Problem," *Journal of Forestry* 20, no. 5 (1922): 437–47.

33 "Causes and Effects of Windfall in Forests," *Pulp and Paper Magazine of Canada*, January 21, 1926, 85–90.

34 Schierbeck pushed the Minister of Lands and Forests, J.C. Douglas, to write to Ottawa to express dismay over what Schierbeck saw as the dominion government's refusal to take meaningful

action to combat the insect. Douglas obliged, and an exchange of angry letters between Ottawa and Halifax ensued: NSA, RG20, 774, Beech Trees and Larch, 13 March 1925, J.M. Swaine to F.A. Harrison; NSA, RG20, 746, Budworm, 30 October 1925, O. Schierbeck to H. Cunningham; NSA, RG20, 718, Attorney-General – Insects and Fungi, 8 January 192[6], L.S. McLaine to O. Schierbeck and 11 January 1926, O. Schierbeck to J.C. Douglas; NSA, RG20, 718, Aa-Ak, 11 February 1926, O. Schierbeck to [dominion] Minister of Agriculture and 22 February 1926, J.H. Grisdale to O. Schierbeck. The war of words escalated when Schierbeck took a swipe at the quality of the dominion's entomologists, and Swaine's ripostes emphatically defended their skills, a subject which is covered in the preceding paragraph.

35 NSA, RG20, 746, Budworm, 10 and 15 December 1925, O. Schierbeck to J.C. Douglas.

36 "Government's New Forestry Chief Named – Otto Schierbeck Head Forester," *Halifax Herald*, March 24, 1926. Creighton insists that Barnjum exercised significant influence over Edgar N. Rhodes, the province's premier, to get Schierbeck appointed Chief Forester: Creighton, *Forestkeeping*, 30.

37 NSA, RG20, 720, Y, 20 February, 5 March and 1 April 1926, J.M. Swaine to O. Schierbeck. Schierbeck tried to backpedal by insisting that his comments were not intended as criticisms of Swaine's personnel, but his denials were disingenuous: NSA, RG20, 720, Y, 8 March 1926, O. Schierbeck to J.M. Swaine. Soon after Schierbeck's article appeared in the *Pulp and Paper Magazine of Canada*, Swaine published a rebuttal to Schierbeck in the country's leading forestry journal: "The Present Status of the Spruce Budworm Outbreaks in Eastern Canada," *Forestry Chronicle* 2, no. 3 (September 1926): 34–6.

38 LAC, RG17, 3047, 40–6(1), 23 December 1925, J.M. Swaine to J.C. Douglas.

39 NSA, RG20, 718, As-4–6, 15 December 1925, O. Schierbeck to J.C. Douglas.

40 NSA, RG20, 718, As-4–6, "Report on Spruce Budworm Conditions in Cape Breton Island, December 1925, M.B. Dunn." Once again, Schierbeck was convinced that he smelled a rat. He wrote to Douglas to express his disbelief upon learning of the views Dunn expressed in the latter's report. Schierbeck explained that he had spoken with Dunn after he had visited Cape Breton and at that time Dunn had been "absolutely in favour of the proposed cutting operations." Schierbeck was thus shocked to learn that Dunn had "arrived at another opinion" upon Dunn's return to Ottawa: NSA, RG20, 718, As-4–5, 4 January 1926, O. Schierbeck to J.C. Douglas.

41 Both Schierbeck and Oxford Paper were now eager to cooperate in publicizing the importance of forestry issues to Nova Scotians: NSA, RG20, 720, Me-Mh, 15 April 1926, G.S. Harvey to O. Schierbeck and 23 April 1926, O. Schierbeck to G.S. Harvey; NSA, RG20, 720, Ra-Rd, 9 August 1926, J.L. Ralston to O. Schierbeck and 12 August 1926, O. Schierbeck to J.L. Ralston.

42 Schierbeck actually wrote the articles that the *Herald* published on its front page about the need for better forest protection: for example, see "Forest Protection Vital to Continued Industrial Welfare," *Halifax Herald*, April 19, 1926.

43 Swaine, "The Present Status of the Spruce Budworm Outbreaks," 34–6.

44 NSA, RG20, 720, Y, 5 March 1926, J.M. Swaine to O. Schierbeck.

45 NSA, RG20, 720, Y, 18 November 1926, J.M. Swaine to O. Schierbeck

46 Forest History Society Archives (FHSA), F16.1 – USFS History Collection, Biographical information on Ernst J. Schreiner; Jacques Cattell Press, *American Men and Women of Science*, 5606; United States Department of Agriculture, *Trees*, 159.

47 NSA, RG20, 720, Y, 29 April and 17 and 27 May 1926, J.M. Swaine to O. Schierbeck; 28 May 1926, J.C. Douglas to J.M. Swaine; 31 May 1926, H.L. Fenerty to J.M. Swaine; 2 June 1926, J.M. Swaine to J.C. Douglas and O. Schierbeck; and 5 June 1926, J.M. Swaine to O. Schierbeck; NSA, RG20, 718, Attorney-General – Miscellaneous 1926, 25 June 1926, O. Schierbeck to J.C. Douglas, from which the citations are taken.

48 NSA, RG20, 719, McA-McC, 20 and 29 June 1926, A.H. MacAndrews to O. Schierbeck and 2 and 13 July 1926, O. Schierbeck to A.H. MacAndrews. Locals were also noticing the budworm infestation on the eastern Nova Scotian mainland and calling for the government to take action against it: NSA, RG20, 719, McA-McC, 2 July 1926, H.F. Macdonald to O. Schierbeck and 1 December 1926, O. Schierbeck to H.F. Macdonald. Initial concerns that the caterpillars on the mainland were not spruce budworm were soon put to rest: NSA, RG20, 720, Y, 15 and 21 July 1926, J.M. Swaine to O. Schierbeck; 21 July 1926, O. Schierbeck to J.M. Swaine; and 28 July 1926, J.M. Swaine to O. Schierbeck.

49 NSA, RG20, 719, McA-McC, 28 and 31 July 1926, J.M. Swaine to O. Schierbeck. The King–Byng Affair was a political crisis that occurred in mid-1926 in Canadian federal politics. It saw the reigning Prime Minister, W.L. Mackenzie King, request a dissolution of Parliament from Lord Byng, the Governor General, in an effort to avoid being censured in the House of Commons for his government's involvement in a customs scandal. Byng refused, and asked the leader of the Opposition to form a government instead. In the ensuing election, King made a stinging rebuke of Byng for having interfered in Canadian politics the central plank in his platform, thereby enabling him to dodge around the customs scandal and win the contest. For details, see Graham, *The King-Byng Affair*.

50 NSA, RG20, 720, Y, 23 November 1926, O. Schierbeck to J.M. Swaine and 3 December 1926, J.M. Swaine to O. Schierbeck.

51 Benjamin and Renlund, *Insecticide Use in Wisconsin Natural Forests and Plantations*.

52 Ibid.; United States Office of Experiment Stations, *Report of the Agricultural Experiment Stations*, 52; Fracker and Granovsky, "The Control of the Hemlock Spanworm by Airplane Dusting"; Fracker and Granovsky, "Airplane Dusting to Control the Hemlock Spanworm." Tellingly, the booklet the Washington Forest Protection Association (WFPA) published to celebrate its centennial history in 2008 described the dusting project it conducted in 1931 as "the first aerial effort to fight a forest pest epidemic in the United States": Washington Forest Protection Association, *A Chronology of the First 100 Years, 7*. I am indebted to Ms. Cindy Mitchell, the WFPA's Senior Director of Public Affairs, for forwarding a copy of this document to me.

53 NSA, RG20, 720, Y, 3 December 1926, J.M. Swaine to O. Schierbeck, from which the citations are taken; see also NSA, RG20, 20, Y, 22 December 1926, J.M. Swaine to O. Schierbeck, enclosing "Progress Report of the Spruce Budworm Investigation in Cape Breton."

54 "Airplanes to Combat Forest Pests," *Pulp and Paper Magazine of Canada*, November 18, 1926, 1415. Germany aggressively tested chemical weapons after the First World War under the aegis of conducting entomological experiments: see Jansen, "Chemical Warfare Techniques for Insect Control."

55 NSA, RG20, 720, Y, 21 December 1926, O. Schierbeck to J.M. Swaine and 22 December 1926, J.M. Swaine to O. Schierbeck [2]; Nova Scotia, *Report of the Department of Lands and Forests, 1926* (Halifax: King's Printer, 1927), 49–50.

56 NSA, RG20, 720, Me-Mh, 15 April 1926, G.S. Harvey to O. Schierbeck and 23 April 1926, O. Schierbeck to G.S. Harvey; NSA, RG20, 720, Ra-Rd, 9 August 1926, J.L. Ralston to O. Schierbeck and 12 August 1926, O. Schierbeck to J.L. Ralston. A short while earlier, Schierbeck had begun developing a very friendly rapport with Ralston's legal partner: see NSA RG20, 720, unlabeled, 30 November 1925, C.J. Burchell to O. Schierbeck.

57 "The Hon. James Layton Ralston," PARLINFO, accessed 26 September 2016, https://lop.parl.ca /sites/ParlInfo/default/en_CA/People/Profile?personId=8357.

58 LAC, RG24-E-1-a, 4912, DND – Operations for Department of Agriculture, Vol. 1, 10 January 1927, H.E. Beedy to J.L. Ralston.

59 LAC, RG24-E-1-a, 4912, DND – Operations for Department of Agriculture, Vol. 1,
    14 January 1927, J.L. Ralston to H.E. Beedy and 14 January 1927, [J.L. Ralston] to Major
    MacDowell.

60 LAC, RG24-E-1-a, 4912, DND – Operations for Department of Agriculture, Vol. 1, 2 February
    1927, T.W. Macdowell to Deputy Minister, with 3 February 1927 note from Desbarats; LAC,
    RG24-E-1-a, 4912, DND – Operations for Department of Agriculture, Vol. 1, 5 February 1927,
    J.H. MacBrien to RCAF.

61 LAC, RG24-E-1-a, 4912, DND – Operations for Department of Agriculture, Vol. 1, 3 February
    1927, E. Wilson to J.M. Swaine. In case the Air Board opted to hire a Canadian firm, Swaine
    had already laid the groundwork for doing so. He had arranged with Ellwood Wilson, one of
    Canada's iconic forestry pioneers and president of Fairchild Aerial Surveys Company in Quebec,
    to fly the mission for a fee of $5,000.

62 LAC, RG24-E-1-a, 4912, DND – Operations for Department of Agriculture, Vol. 1, 20 January
    1927, J.M. Swaine, Memorandum on proposed airplane dusting; see also 21 January 1927,
    A. Gibson to J.H. Grisdale and 27 January 1927, G.J. Desbarats to J.L. Ralston; LAC, RG17,
    3131, 69, 2 February 1927, J.H. Grisdale to G.J. Desbarats.

63 LAC, RG24-E-1-a, 4912, DND – Operations for Department of Agriculture, Vol. 1, 24 January
    1927, H.E. Beedy to J.M. Swaine; 2 February 1927, Department of Agriculture to Deputy
    Minister, DND; and 2 February 1927, T.W. Macdowell to Deputy Minister, DND.

64 LAC, RG24-E-1-a, 4912, DND – Operations for Department of Agriculture, Vol. 1, 7 February
    1927, J.S. Scott to J.H. MacBrien; 10 February 1927, J.H. MacBrien to T.W. Macdowell; and 23
    February 1927, J.H. MacBrien to J.S. Scott, from which the citation is taken. The RCAF was
    fervently committed to generating publicity about the civilian work it was doing, and the media
    obliged by publishing endless stories about its activities in this regard: see, for example, "Aviation
    in Canada Is Not Spectacular, But Highly Efficient," *Globe*, June 10, 1927; "Civil Aviation in
    Dominion," *Globe*, September 4, 1928.

65 LAC, RG24-E-1-a, 4912, DND – Operations for Department of Agriculture, Vol. 1, 25 February
    1927, J.S. Scott to J.H. MacBrien. It was hardly a coincidence that, in the meantime, offers to
    carry out the work arrived on practically a weekly basis from aerial contracting firms in both the
    US and Canada: see LAC, RG24-E-1-a, 4912, DND – Operations for Department of Agriculture,
    Vol. 1, 24 January 1927, A.S. Morse to J.M. Swaine; 3 February 1927, E. Wilson to J.M. Swaine;
    and 8 February 1927, C.E. Woolman to J.M. Swaine.

66 LAC, RG24-E-1-a, 4912, DND – Operations for Department of Agriculture, Vol. 1, 2 March
    1927, J.S. Scott to J.H. MacBrien, from which the citation is taken and on which MacBrien
    has written his comments to Scott; see also 28 February 1927, MacBrien, Memorandum re:
    budworm outbreak.

67 LAC, RG24-E-1-a, 4912, DND – Operations for Department of Agriculture, Vol. 1, 18 April
    1927, B.R. Coad to E. Wilson.

68 LAC, RG24-E-1-a, 4912, DND – Operations for Department of Agriculture, Vol. 2, 25 April
    1927, B.R. Coad to E. Wilson. See also 25 March 1927, B.R. Coad to J.M. Swaine; 11 April
    1927, E. Wilson to B.R. Coad; LAC, RG24-E-1-a, 4912, DND – Operations for Department
    of Agriculture, Vol.1, 8 March 1927, B.R. Coad to J.M. Swaine. Douglas explains that Shields
    and Bath spent time in Tallulah but does not mention the deficiencies of their training there:
    Douglas, *The Creation of the National Air Force*, 115–16.

69 LAC, RG24-E-1-a, 4912, DND – Operations for Department of Agriculture, Vol. 2,
    correspondence regarding training the pilots and flying the two Puffers from Pennsylvania to
    Manitoba and Nova Scotia between early May and late June 1927, 25 April 1927, B.R. Coad to
    J.M. Swaine.

70 LAC, RG24-E-1-a, 4912, DND – Operations for Department of Agriculture, Vol. 2, correspondence regarding training the pilots and flying the two Puffers from Pennsylvania to Manitoba and Nova Scotia between early May and late June 1927, 4 June 1927, RCAF to Keystone Corporation.

71 LAC, RG24-E-1-a, 4912, DND – Operations for Department of Agriculture, Vol. 2, correspondence regarding training the pilots and flying the two Puffers from Pennsylvania to Manitoba and Nova Scotia between early May and late June 1927, C.L. Bath, Report on Huff-Daland Duster Seaplane Equipped with Wright J.5 Engine. See also NSA, RG20, 721, Su-Sz, 1, 4 and 22 June, and 19 [2] and 23 July 1927, J.M. Swaine to O. Schierbeck; 29 June 1927, United Fruit to Nova Scotia Department of Lands and Forests; and 2 July 1927, C.L. Bath to O. Schierbeck.

72 "Forester To Use Aeroplane: Novel Plan To Be Tried in Cape Breton Next Month," *Halifax Herald*, May 23, 1927.

73 "Spray and Dust Must Be Kept Up," *Halifax Herald*, June 4, 1927; "Urge Move To Assist Growers," *Halifax Herald*, June 14, 1927; "Fruit Growers To Form 'Spray Circles,'" *Halifax Herald*, June 25, 1927; "Horticultural Experts Will Demonstrate New Methods of Spraying," *Halifax Herald*, July 15, 1927.

74 "To Aid in the Protection of Forests," *Halifax Herald*, June 14, 1927; "Brings New Plane," *Montreal Gazette*, June 14, 1927. The plane's progress can be tracked through the following correspondence: LAC, RG24-E-1-a, 4912, DND – Operations for Department of Agriculture, Vol. 2, 10 June 1927, C.L. Bath to Secretary RCAF [telegram]; 12 June 1927, Keystone Corporation to Secretary RCAF [telegram]; and 13 and 14 June 1927, DND Dartmouth to Secretary, DND [telegram].

75 LAC, RG24-E-1-a, 4912, DND – Operations for Department of Agriculture, Vol. 2, 30 May 1927, HQ Order No. K4. When Flying Officer Bath was asked to explain the nature of the work that he was doing in the province, he only ever mentioned being based at Whycocomagh and dusting over Cape Breton: "Begin Now To Fight All Pests," *Halifax Herald*, May 28, 1927.

76 NSA, RG20, 721, McA-McC, 17 May 1927, C.N. Pertus to O. Schierbeck, 22 February 1927, J.M. Swaine to O. Schierbeck; 1 March and 10, 17 and 18 May 1927, O. Schierbeck to J.M. Swaine; and 9 and 16 May 1927, J.M. Swaine to O. Schierbeck; NSA, RG20, 721, Su-Sz, 5 April 1928, J.M. Swaine to O. Schierbeck, enclosing "A Summary Report on an Airplane Dusting Experiment for Control of the Spruce Budworm Conducted in Cape Breton Island, June 1927, J.M. Swaine." Swaine published the identical document in Canada Department of National Defence, *Report on Civil Aviation and Civil Government Air Operations for the Year 1928* (Ottawa: F.A. Acland, Printer to the King's Most Excellent Majesty, 1929). Both reports stated that a final report would be forthcoming, but it never appeared.

77 Canada Department of Agriculture, *Annual Report of the Minister of Agriculture for the Dominion of Canada for the Year Ended March 31, 1928* (Ottawa: F.A. Acland, Printer to the King's Most Excellent Majesty, 1928), 107.

78 NSA, RG20, 721, Su-Sz, 17 August 1927, J.J. de Gryse to O. Schierbeck, enclosing "Report of the Spruce Budworm Dusting in Cape Breton Island [dictated but not read by J.M. Swaine]."

79 Ibid.

80 Like Oxford Paper, Westvaco was a major player in the American pulp and paper industry: CUA – Division of Rare and Manuscript Collections, MSS1781, 108, 53–54, Oxford Paper Company, all correspondence.

81 CUA – Division of Rare and Manuscript Collections, MSS1781, 26, 1–3, Sonora Timber Company 1924–1926, documents dealing with the firm's timber holdings in Nova Scotia; NSA, RG20, 718, As-4–6, 5 November 1925, O. Schierbeck to J.C. Douglas; Legge, *Sawdust and Sea Breezes*.

82 CUA – Division of Rare and Manuscript Collections, MSS1781, 26, 1–3, Sonora Timber Company 1924–1926, documents dealing with the firm's timber holdings in Nova Scotia; Cahill, *The Thousandth Man*; *Moody's Manual of Industrials, 1920s*, entries for "Oxford Paper Company" and "West Virginia Pulp and Paper Company."

83 Westvaco had asked for a reduction in the dues it paid on the budworm-damaged timber it harvested, but it is unclear if the Nova Scotia government granted the request. It is possible that, instead, the government agreed to cover the cost of the dusting operation over Westvaco's timber holdings; see NSA, RG20, 721, Sn-So, 10 November 1926, T.N. Agapayeff to W.L. Hall.

84 LAC, RG24-E-1-a, 4912, DND – Operations for Department of Agriculture, Vol. 2, 24 June 1927, RCAF to Officer Commanding, Dartmouth, NS; see also 24 June 1927, J.M. Swaine to J.A. Wilson and 24 June 1927, RCAF to J.M. Swaine. The dominion entomology branch barely mentioned in its annual report the operation in Guysborough County, and even misidentified where it occurred. It explained that it had had no officials directly involved in the effort, and that it was the Nova Scotia Forest Branch (i.e., Schierbeck) that provided "a large amount of insecticide for a continuation of the experiment on a budworm outbreak in Antigonish county [*sic*]": Canada Department of Agriculture, *Annual Report of the Minister of Agriculture for the Dominion of Canada for the Year Ended March 31, 1928*, 107. Similarly, when the *Pulp and Paper Magazine of Canada* reported on the experimental dusting in Nova Scotia, it described in significant detail the project that occurred on Cape Breton Island but was silent about the one in Guysborough County: "War on Forest Pests Started," *Pulp and Paper Magazine of Canada*, August 4, 1927.

85 Nova Scotia, *Report of the Department of Lands and Forests, 1927* (Halifax: King's Printer, 1928), 31.

86 NSA, RG20, 742, Barnjum Geo.W. + F.J.D., 12 July 1927, O. Schierbeck to F.J.D. Barnjum.

87 LAC, RG24-E-1-a, 4912, DND – Operations for Department of Agriculture, Vol. 2, 28 July 1927, Aeronautics [Dartmouth] to Secretary, DND; NSA, RG20, 721, Su-Sz, 22 June 1927, O. Schierbeck to J.M. Swaine and 5 January 1928, J.M. Swaine to O. Schierbeck; NSA, RG20, 721, Sn-So, 7 July 1927, Sonora Timber Company to Nova Scotia Department of Forestry; NSA, RG20, 721, Su-Sz, 17 August 1927, J.J. de Gryse to O. Schierbeck, enclosing "Report of the Spruce Budworm Dusting in Cape Breton Island [dictated but not read by J.M. Swaine]."

88 LAC, RG24-E-1-a, 4912, DND – Operations for Department of Agriculture, Vol. 2, 2 August 1927, J.M. Swaine to J.A. Wilson.

89 J.M. Swaine, "Progress in Forest Insect Control in Canada," *Forestry Chronicle* 4, no. 1 (1928): 35–40.

90 LAC, RG24-E-1-a, 4912, DND – Operations for Department of Agriculture, Vol. 2, 2 August 1927, E. Huff to Secretary, DND, with enclosures including "War on Forest Pests Dusting Trees by Aeroplane," *London Times*, July 25, 1927. See also LAC, RG24-E-1-a, 4912, DND – Operations for Department of Agriculture, Vol. 2, 30 May 1927, W.S. Brancker to J.A. Wilson; 13 June 1927, J.A. Wilson to W.S. Brancker; and 25 July 1927, J.A. Wilson to RCAF Liaison Officer, Air Ministry, Britain.

91 LAC, RG24-E-1-a, 4912, DND – Operations for Department of Agriculture, Vol. 2, 17 June 1927, T.G. Hetherington to Director, RCAF; NSA, RG20, 724, Pe-Ph, 6 April 1928, H.B. Peirson to O. Schierbeck and 17 April 1928, O. Schierbeck to H.B. Peirson.

92 LAC, RG24-E-1-a, 4899, 1008-2-2, Vol. 5, 28 April 1928, W.W. Cory to G.J. Desbarats. Cory's department also published very positive accounts of the operation long before its results were even understood: Canada Department of the Interior, *Natural Resources Canada*. Officials in Nova Scotia also boasted to their Australian audience that their dusting trials in 1927 had been

a smashing success: NSA, O/S V/F 28-10, *Forests and Forestry in Nova Scotia, W.L. Hall and Otto Schierbeck – Special Report to the British Empire Forestry Conference, Australia 1928.*

93 LAC, RG24-E-1-a, 4912, DND – Operations for Department of Agriculture, Vol. 2, 25 July 1927, J.A. Wilson to RCAF Liaison Officer, British Air Ministry. Wilson's report was apparently the source for the article that appeared in the London press, and conspicuously absent from it was an assessment of the project's success: "War on Forest Pests Dusting Trees by Aeroplane," *London Times*, July 25, 1927.

94 "War on Forest Pests Started," *Pulp and Paper Magazine of Canada*, August 4, 1927.

95 NSA, RG20, 724, Su-Sz, 5 April 1928, J.M. Swaine to O. Schierbeck, enclosing "A Summary Report on an Airplane Dusting Experiment for Control of the Spruce Budworm Conducted in Cape Breton Island, June 1927, J.M. Swaine."

96 NSA, RG20, 721, Su-Sz, 19 July 1927, J.M. Swaine to O. Schierbeck. In Swaine's final report on the effort in Nova Scotia, he devoted considerable attention to outlining the practical difficulties encountered in trying to count the number of dead caterpillars: NSA, RG20, 724, Su-Sz, 5 April 1928, J.M. Swaine to O. Schierbeck, enclosing "A Summary Report on an Airplane Dusting Experiment for Control of the Spruce Budworm Conducted in Cape Breton Island, June 1927, J.M. Swaine."

97 NSA, RG20, 721, Su-Sz, 27 July 1927, O. Schierbeck to J.M. Swaine; NSA, RG20, 724, Su-Sz, 5 January 1928, J.M. Swaine to O. Schierbeck, from which the citation is taken.

98 NSA, RG20, 724, Su-Sz, 5 April 1928, J.M. Swaine to O. Schierbeck, enclosing "A Summary Report on an Airplane Dusting Experiment for Control of the Spruce Budworm Conducted in Cape Breton Island, June 1927, J.M. Swaine." Swaine republished much of the same report in the *Pulp and Paper Magazine of Canada* on 12 April 1928.

99 Ibid.

100 Ibid.

101 NSA, RG20, 723, Du-Dz, 4 September 1928, J.A. Duchastel to O. Schierbeck; 11 September 1928, O. Schierbeck to J.A. Duchastel, from which the first citation is taken; and 13 September 1928, J.A. Duchastel to O. Schierbeck, from which the second citation is taken.

102 NSA, RG20, 724, Su-Sz, 5 April 1928, J.M. Swaine to O. Schierbeck, enclosing "A Summary Report on an Airplane Dusting Experiment for Control of the Spruce Budworm Conducted in Cape Breton Island, June 1927, J.M. Swaine."

3 "Fighting insect plagues is something new"

1 University of Toronto Archives (UTA), A1972–0025, 143, unlabeled file, 7 December 1920, C.D. Howe to J.E. Rothery; UTA, A1972–0025, 149, Wia–Wil, 8 February 1922, C.D. Howe to E. Wilson.

2 Kuhlberg, "'We are the pioneers in this business.'"

3 St. Mary's Paper Archives (SMPA), A-1, Forestry 1921, August 1921, B.F. Avery to G.H. Gray, enclosing [undated] "Working Plan for Goulais Watershed, 13."

4 SMPA, A-1, Forestry 1923, 31 March 1923, B.F. Avery to G.H. Gray, enclosing "4th Annual Report of the Forestry Branch, B. F. Avery," 21

5 SMPA, A-1, Forestry 1925, 1 March 1925, B.F Avery to G.H. Gray, enclosing "6th Annual Report of the Forestry Branch, B.F. Avery," passim; Abitibi-Bowater Archives – Iroquois Falls Division, Forest History Files, January 1925, Plan of Reforestation, Woods Department, Engineering Division, Abitibi Power & Paper Company, 15–16: Abitibi's officials incorrectly assumed that white spruce was immune to the budworm, whereas it was simply less vulnerable than balsam

fir because its buds opened a little bit later in the spring than the former species. Castonguay, *Protection des cultures*, 138, asserts that Canada's pulp and paper industry did not consider it a high priority to alter its harvesting techniques in an effort to combat forest insects, but the evidence suggests otherwise, at least for much of the 1920s.

6 SMPA, A-1, Forestry 1927, September 1927, Report on Aerial Forest Survey – portions of Spanish and Sturgeon Concessions for Spanish River Pulp & Paper Mills Limited, James D. Lacey & Co. (Canada) Limited.

7 Canada Department of Agriculture, Entomological Branch, *The Canadian Insect Pest Review* 5, no. 2 (30 June 1927): 19; and 5, no. 3 (31 July 1927): 32.

8 AO, RG1–256, Box 1, Spruce Budworm, 24 February 1928, B.F. Avery to E.J. Zavitz.

9 AO, RG1–256, Box 1, Spruce Budworm, 15 February 1928, B.F. Avery to E.J. Zavitz; see also 24 February 1928, B.F. Avery to E.J. Zavitz.

10 AO, RG1–256, Box 1, Spruce Budworm, 13 and 14 February 1928, K.A. Stewart to E.J. Zavitz.

11 AO, RG1–256, Box 1, Spruce Budworm, 8 March 1928, J.M. Swaine to E.J. Zavitz.

12 LAC, RG39, 341, 47478, 4 April 1928, E.H. Finlayson to Col. J.L. Gordon; AO, RG1–256, Box 1, Spruce Budworm, 10 April 1928, W.R. Maxwell to E.J. Zavitz.

13 LAC, RG24-E-1-a, 4912, DND – Operations for Department of Agriculture, Vol. 2, 29 April 1927, C.L. Bath, Report on "Dusting Courses"; see also 29 April 1927, T.M. Shields to CO, Borden.

14 LAC, RG24-E-1-a, 4912, DND – Operations for Department of Agriculture, Vol. 1, 4 March 1927, C.T. Porter to E.W. Stedman.

15 LAC, RG24-E-1-a, 4912, DND – Operations for Department of Agriculture, Vol. 1, 11 March 1927, C.T. Porter to E.W. Stedman.

16 LAC, RG24-E-1-a, 4912, DND – Operations for Department of Agriculture, Vol. 1, 21 February 1927, J.M. Swaine to J.A. Wilson; 4 March 1927, E.W. Stedman to Huff-Daland Airplanes Inc.; LAC, RG24-E-1-a, 4912, DND – Operations for Department of Agriculture, Vol. 2, 12 March 1927, E.W. Stedman to Director of Contracts, DND; 15 March 1927, J.A. McCann to Keystone Aircraft Corporation; and 23 March 1927, Keystone Aircraft Corporation to RCAF [telegram].

17 LAC, RG24-E-1-a, 4912, DND – Operations for Department of Agriculture, Vol. 2, 7 July 1927, Supreme Officer Commanding, Report on Huff-Daland; and 30 July 1927, C.L. Bath, Report on Huff Daland, from which the citations are taken. Compounding the Air Board's concerns about the Puffer were its cost and lack of versatility compared to other aircraft. Over the summer and fall of 1927, the Air Board communicated with the Dominion Forest Branch about which aircraft would be best to order for use in future forestry work. All agreed that DH Moths were indisputably better suited to this task than the Puffer. Not only were they able to fly at higher ceilings, which was needed for forest fire patrol work, but four Moths could be obtained for the price of one Puffer, and the latter were twice as expensive to maintain: LAC, RG24-E-1-a, 4899, 1008–2-2, Vols. 4–5, 12 and 27 August and 1 October 1927, J.L. Gordon to E.H. Finlayson; 19 August and 6 October 1927, E.H. Finlayson to J.L. Gordon; and 25 August 1927, J.L. Gordon to G.J. Desbarats.

18 LAC, RG23-E-1-a, 4912, DND – Operations for Department of Agriculture, Vol. 3, 28 March 1928, C.L. Bath to Director, Civil Government Air Operations.

19 Ibid.

20 LAC, RG23-E-1-a, 4912, DND – Operations for Department of Agriculture, Vol. 3, 28 March 1928, C.L. Bath to Director, Civil Government Air Operations, on which Gordon and Stedman have recorded their thoughts.

21 In contrast, the Puffer had certainly proven in 1927 to be well suited to dumping poisons over the Canadian Prairies in an effort to combat wheat rust, and this caused officials to request the

same plane and pilot be used again the following season: LAC, RG17, 3131, 69 – Department of Militia and Defence, 1 April 1928, E.S. Archibald to J.H. Grisdale and 10 April 1928, G.J. Desbarats to J.H. Grisdale.

22 "Aviation's Many Uses Described," *Montreal Gazette*, February 11, 1928.

23 LAC, RG24-E-1-a, 4912, DND – Operations for Department of Agriculture, Vol. 3, 5 April 1928, J.L. Gordon to Deputy Minister.

24 LAC, RG24-E-1-a, 4912, DND – Operations for Department of Agriculture, Vol. 3, 13 April 1928, G.J. Desbarats to J.H. Grisdale.

25 LAC, RG39, 341, 47478, 10 April 1928, J. L. Gordon to E. H. Finlayson.

26 LAC, RG17, Vol. 3131, 69 – Department of Militia and Defence, 26 April 1928, G.J. Desbarats to J.H. Grisdale; see also 14 April 1928, A. Gibson to J.H. Grisdale and 19 April 1928, J.H. Grisdale to G.J. Desbarats; LAC, RG39, 341, 47478, 30 April 1928, J.H. Grisdale to G.J. Desbarats.

27 SMPA, A-1, Forestry 1928, Report of Reconnaissance Flight made for purpose of locating sample plots for experimental dusting of budworm infested timber, enclosed in 15 May 1928, A.H. Burk to J.M. Swaine.

28 SMPA, A-1, Forestry 1928, 15 May 1928, Instructions for Spruce Budworm Plots; and Report of Ground investigation of areas located from the air as possible plots for experimental dusting of Budworm infested timber, enclosed in 22 May 1928, A.H. Burk to J.M. Swaine.

29 SMPA, A-1, Forestry 1928, 16 May 1928, J.M. Swaine to A.H. Burk.

30 SMPA, A-1, Forestry 1928, Report of Ground investigation of areas located from the air as possible plots for experimental dusting of Budworm infested timber, enclosed in 22 May 1928, A.H. Burk to J.M. Swaine; see also J.M. Swaine, "Forest Insect Investigations in Canada 1928," *Forestry Chronicle* 5, no. 2 (Spring 1929).

31 AO, RG1–256, Box 1, Spruce Budworm, 4 June 1928, K.A. Stewart to E.J. Zavitz; 11 June 1928, C.R. Mills to E.J. Zavitz; and 14 June 1928, K.A. Stewart to E.J. Zavitz, enclosing "Progress Report – Spruce Budworm Control Experiment"; LAC, RG24-E-1-a, 4912, DND – Operations for Department of Agriculture, Vol. 3, 4 May 1928, E.J. Zavitz to J.A. Wilson; 21 May and 1 June 1928, J.M. Swaine to J.L. Gordon; and 5 June 1928, Headquarters' Order AC 46/28/Dusting.

32 AO, RG1–256, Box 1, Spruce Budworm, 6 June 1928, J.M. Swaine to E.J. Zavitz.

33 Canada Department of National Defence, *Report on Civil Aviation and Civil Government Air Operations for the Year 1928* (Ottawa: F.A. Acland, Printer to the King's Most Excellent Majesty, 1929), Appendix D. In contrast, the report described in significant detail the highly successful aerial dusting trial that the Air Board conducted against wheat rust in Manitoba that season.

34 J.M. Swaine, "Forest Insect Investigations in Canada 1928," *Forestry Chronicle* 5, no. 2 (Spring 1929), from which the first citation is taken; AO, RG1–256, Box 1, Spruce Budworm, 29 June 1928, K.A. Stewart to E.J. Zavitz, from which the other citations are taken.

35 SMPA, A-1, Forestry – 1928, 18 July 1928, A.H. Burk to B.F. Avery. See also SMPA, A-1, Forestry – 1928, "Outline of the Westree Budworm Investigation by A.H. MacAndrews, July 7/28"; AO, RG1–256, Box 1, Spruce Budworm, 4 July 1928, K.A. Stewart to E.J. Zavitz [2]; J.M. Swaine, "Forest Insect Investigations in Canada 1928," *Forestry Chronicle* 5, no. 2 (Spring 1929): 35–41.

36 SMPA, A-1, Forestry – 1928, 18 July 1928, A.H. Burk to B.F. Avery.

37 Canada Department of Agriculture, Entomological Branch, *The Canadian Insect Pest Review* 6, no. 2 (25 June 1928): 17; and 6, no. 5 (5 October 1928): 46.

38 LAC, RG24-E-1-a, 4912, DND – Operations for Department of Agriculture, Vol. 3, 5 July 1928, C.L. Bath, Memorandum, and 6–7 July 1928, R.S. Grandy to Secretary, DND, from which all the citations are taken. Notably, the Air Board's aversion to participating in these missions going forward did not prevent it from publicizing in one of the world's leading aeronautical journals the groundbreaking aerial forest dusting work it had conducted in 1927 and 1928 against the

budworm. In fact, it said nothing about the dangers involved in this type of flying and depicted itself as having saved the day by carrying out the missions: *Flight Magazine*, August 16, 1928, 713 and September 20, 1928, 808–9.

39 LAC, RG24-E-1-a, 4912, DND – Operations for Department of Agriculture, Vol. 3, 11 July 1928, J.M. Swaine to J.L. Gordon.

40 AO, RG1–256, Box 1, Spruce Budworm, 18 July 1928, J.M. Swaine to K.A. Stewart.

41 LAC, RG17, 3131, 69 – Department of Militia and Defence, 19 November 1928, G.J. Desbarats to J.H. Grisdale.

42 Kuhlberg, "'We are the pioneers in this business,'" passim.

43 LAC, RG17, 3131, 69 – Department of Militia and Defence, 21 November 1928, A.T. Charron to H.T. Gussow and A. Gibson; 27 November 1928, A. Gibson to A.T. Charron; and 29 November 1928, H.T. Gussow to A.T. Charron.

44 Swaine, *The Hemlock Looper*; "Hemlock Looper," National Resources Canada, accessed 18 November 2016, https://tidcf.nrcan.gc.ca/en/insects/factsheet/8846.

45 Wiegman, *Trees to News*, passim.

46 Ibid., 118–20; Canada Department of Agriculture, Entomological Branch, *The Canadian Insect Pest Review* 6, no. 5 (5 October 1928): 46; NSA, RG20, 723, Du-Dz, 4 September 1928, J.A. Duchastel to O. Schierbeck; 11 September 1928, O. Schierbeck to J.A. Duchastel; and 13 September 1928, J.A. Duchastel to O. Schierbeck; LAC, RG24-E-1-a, 4913, DND – Operations for Department of Agriculture, Vol. 4, "Preliminary Report Airplane Dusting – Manicouagan 1929," J.M. Swaine; LAC, RG17, 3131, 69 – Department of Militia and Defence, 21 November 1928, A.T. Charron to H.T. Gussow and A. Gibson; 27 November 1928, A. Gibson to A.T. Charron; and 29 November 1928, H.T. Gussow to A.T. Charron.

47 AO, RG1–256, Box 1, Spruce Budworm, 9 February 1929, J.M. Swaine to E.J. Zavitz; see also 25 March 1929, E.J. Zavitz to J.M. Swaine.

48 LAC, RG24-E-1-a, 4913, DND – Operations for Department of Agriculture, Vol. 4, 8 April 1929, W. Knight (for Junkers Corporation of America) to J.M. Swaine and 10 April 1929, J.M. Swaine to J.L. Gordon; LAC, RG17, 3131, 69 – Department of Militia and Defence, 2 May 1929, G.J. Desbarats to J.H. Grisdale; Natural Resources Canada – Pacific Forestry Centre Archives (PFCA), Project Western Hemlock Looper 1929–1930, 18 December 1929, J.M. Swaine to G. Hopping; Pigott, *Taming the Skies*, 47.

49 *Mail and Empire*, January 29, 1929; *Border Cities Star*, June 1, 1929.

50 LAC, RG24-E-1-a, 4913, DND – Operations for Department of Agriculture, Vol. 4, 1 June 1929, K. Kodama to Department of Agriculture.

51 The film is preserved in two parts and contains some truly remarkable footage: NFB, *An aeroplane dusting experiment for the control of spruce bud worm carried out at Westree on* [*sic*] *June 1929*. It would be subsequently used to publicize the entomologists' work in public lectures across Canada: AO, RG1–560, 14 February 1931, *Fredericton Gleaner*, "Tells of War Being Waged on Insects."

52 AO, RG1–256, Box 1, Spruce Budworm, 31 May and 15 June 1929, K.A. Stewart to J.M. Swaine; 4 and 7 June 1929, K.A. Stewart to J.M. Swaine and E.J. Zavitz; 5 June 1929, J.M. Swaine to E.J. Zavitz; and October 1929, "Spruce Budworm Control Experiment – Season 1929," M.W. Kensit; LAC, RG24-E-1-a, 4913, DND – Operations for Department of Agriculture, Vol. 4, 6 June 1929, HQ Order AC70/29/Dusting.

53 LAC, RG24-E-1-a, 4913, DND – Operations for Department of Agriculture, Vol. 4, 15–23 June 1929, telegrams from N.C. Ogilvie-Forbes to DND with the citations taken from 15 June, 17 June, and 23 June; see also 6 August 1929, N.C. Ogilvie-Forbes to Commanding Officer, Ottawa Air Station.

54 LAC, RG24-E-1-a, 4913, DND – Operations for Department of Agriculture, Vol. 4, 17 April 1930, J.H. Grisdale to J.L. Gordon, attached to which is "Airplane Dusting Operations for the Control of Defoliating Insects – Conducted in Cooperation with the Dominion Air Service in 1929," J.M. Swaine, from which the citation is taken; the identical report was published in Canada Department of National Defence, *Report on Civil Aviation and Civil Government Air Operations for the Year 1929* (Ottawa: F.A. Acland, Printer to the King's Most Excellent Majesty, 1930). See also LAC, RG24-E-1-a, 4913, DND – Operations for Department of Agriculture, Vol. 4, 6 August 1929, N.C. Ogilvie-Forbes to Commanding Officer, Ottawa Air Station.

55 LAC, RG39, 341, 47278 – Head Office – Pests – Dusting Experiments, 13 June 1929, J.M. Swaine to J.R. Dickson.

56 LAC, RG24-E-1-a, 4913, DND – Operations for Department of Agriculture Vol. 4, 17 April 1930, J.H. Grisdale to J.L. Gordon, attached to which is "Airplane Dusting Operations for the Control of Defoliating Insects – Conducted in Cooperation with the Dominion Air Service in 1929," J.M. Swaine; Canada Department of National Defence, *Report on Civil Aviation and Civil Government Air Operations for the Year 1929* (Ottawa: F.A. Acland, Printer to the King's Most Excellent Majesty, 1930).

57 AO, RG1–560, 14 February 1931, *Fredericton Gleaner*, "Tells of War Being Waged on Insects." The *Gleaner* reported on this matter in 1931 after Swaine delivered a public lecture in Fredericton about his campaign against the budworm, which included a screening of the NFB's short documentary about the 1929 Westree operation.

58 SMPA, A-1, Forestry 1932, March 1933, Report of Organization for Forest Protection and Fire Data for the Sudbury Division of the Abitibi Power & Paper Company Limited, 1932, 34.

59 SMPA, A-1, Forestry 1930, 8 December 1930, A.H. Burk to B.F. Avery.

60 SMPA, A-1, Forestry 1932, March 1933, Report of Organization for Forest Protection and Fire Data for the Sudbury Division of the Abitibi Power & Paper Company Limited, 1932, 10.

61 Canada Department of Agriculture, Entomological Branch, *The Canadian Insect Pest Review* 7, no. 3 (15 July 1929): 33. Riegert, *From Arsenic to DDT*, passim (but especially chap. 20), asserts that Swaine was acutely aware that he had to justify the existence of his division, and find grounds for requesting larger appropriations, by demonstrating the effectiveness of his work in general and the aerial dusting in particular. Otherwise, there was little point in the dominion government investing in these activities.

62 LAC, RG24-E-1-a, 4913, DND – Operations for Department of Agriculture, Vol. 4, 10 June 1929, J.M. Swaine to J.L. Gordon.

63 LAC, RG24-E-1-a, 4913, DND – Operations for Department of Agriculture, Vol. 4, 17 June 1929, J.L. Gordon to G.J. Desbarats; 13 June 1929, A. Gibson to J.H. Grisdale; and 14 June 1929, J.H. Grisdale to G.J. Desbarats; Douglas, *The Creation of the National Air Force*, 77.

64 Swaine circumvented his superiors in the Department of Agriculture again in mid-July when he learned that his colleagues in British Columbia sought to use the trimotor Ford for their own dusting experiment. This time the Air Board bluntly rejected Swaine's request. LAC, RG24-E-1-a, 4913, DND – Operations for Department of Agriculture, Vol. 4, 11 July 1929, J.M. Swaine to J.L. Gordon [telegram] and 12 July 1929, Civil Air Operations to J.M. Swaine.

65 LAC, RG24-E-1-a, 4913, DND – Operations for Department of Agriculture, Vol. 4, 30 May 1929, J.A. Duchastel to A.A. Schmon, enclosed in 30 May 1929, J.A. Duchastel to J.L. Gordon; 10 June 1929, A.A. Schmon to J.M. Swaine; 12 June 1929, J.M. Swaine to J.D. Gilmour; and 14 June 1929, J.D. Gilmour to J.M. Swaine.

66 LAC, RG24-E-1-a, 4913, DND – Operations for Department of Agriculture, Vol. 4, 22 June 1929, C.R. Dunlap to DND [telegram], from which the citation is taken; see also LAC, RG17,

3131, 69 – Department of Militia and Defence, 13 June 1929, A. Gibson to J.H. Grisdale and 3 July 1929, G.J. Desbarats to J.H. Grisdale.

67 LAC, RG24-E-1-a, 4913, DND – Operations for Department of Agriculture, Vol. 4, 5 July 1929, N.C. Ogilvie-Forbes to DND and 5–30 July 1928, telegrams from N.C. Ogilvie-Forbes to DND.

68 LAC, RG24-E-1-a, 4913, DND – Operations for Department of Agriculture, Vol. 4, "Preliminary Report Airplane Dusting – Manicouagan 1929," J.M. Swaine; see also LAC, RG24-E-1-a, 4913, DND – Operations for Department of Agriculture, Vol. 4, 6 August 1929, N.C. Ogilvie-Forbes to Commanding Officer, Ottawa Air Station.

69 LAC, RG24-E-1-a, 4913, DND – Operations for Department of Agriculture, Vol. 4, 17 April 1930, J.H. Grisdale to J.L. Gordon, enclosing "Airplane Dusting Operations for the Control of Defoliating Insects – Conducted in Cooperation with the Dominion Air Service in 1929," J.M. Swaine.

70 Canada Department of Agriculture, Entomological Branch, *The Canadian Insect Pest Review* 9, no. 4 (25 August 1931): 88; see also 8, no. 3 (15 July 1928): 39; 7, no. 4 (15 August 1929): 33; 7, no. 5 (1 October 1929): 58; 8, no. 1 (1 May 1930): 10; 8, no. 6 (15 October 1930): 78–9; and 9, no. 1 (1 May 1931): 21.

71 LAC, RG24-E-1-a, 4913, DND – Operations for Department of Agriculture Vol. 4, 17 April 1930, J.H. Grisdale to J.L. Gordon, enclosing "Airplane Dusting Operations for the Control of Defoliating Insects – Conducted in Cooperation with the Dominion Air Service in 1929," J.M. Swaine.

72 LAC, RG24-E-1-a, 4913, DND – Operations for Department of Agriculture Vol. 4, 11 September 1929, J.M. Swaine to J.L. Gordon; see also 23 December 1929, J.A. Duchastel to J.L. Gordon.

73 LAC, RG17, 3037, Destructive Insect and Pest Act – General (5), "Airplane Dusting for the Control of Forest Insects in Canada," reprint from *Canadian Woodlands Review*, ca. fall 1929. A telling example of how the legacy of the operation against the looper in Quebec lived on is found in Gibson, "The Canadian Entomological Service," 1460, which relates how Gibson boasted about the operation at an international entomology conference in Germany on the eve of the Second World War.

4 "For the sake of this beautiful playground"

1 Russell, "Speaking of Annihilation"; Dunlap, *DDT*, chaps. 1 and 2. Carson, *The Sea Around Us*, writes of the degree to which modern urban life sprouted a warped sense of power over nature.

2 Ontario GEOservices Centre and Ministry of Northern Development and Mines, *Rock ONtario*, chap. 5.

3 Drummond, *Progress without Planning*, chap. 5; Lower, *Great Britain's Woodyard*; Lower, *The North American Assault on the Canadian Forest*.

4 McMurray, *The Free Grant Lands of Canada*, 17; *Picturesque Canada: The Northern Lakes Guide*; *Information for Intending Settlers*; Baxter and Hall, "The Muskoka Lakes Highlands of Ontario, Canada," 329; Jasen, *Wild Things*, passim, but especially chap. 5; Hodgins and Benedickson, *The Temagami Experience*.

5 Lundell, *Old Muskoka*; Rogers, *Guide Book and Atlas of Muskoka and Parry Sound Districts*, 1–13.

6 Rogers, *Muskoka Lakes Blue Book*; Brown and Cook, *Canada*, 211.

7 *Muskoka: Land of Health and Pleasure*.

8 Baxter and Hall, "The Muskoka Lakes Highlands of Ontario, Canada," 329.

9 Ibid., from which the citation is taken; Adam, *Muskoka Illustrated*; Benedickson, *The Culture of Flushing*, 163.

10 Leather-makers cherished hemlock's bark for tanning, and throughout the second half of the nineteenth and early twentieth centuries they destroyed groves of it solely to acquire its trunk's outermost layer. In addition, its reputation for being naturally resistant to rotting made it highly sought after when Toronto built its subway in the early 1950s.

11 AO, RG1–256, 1, Hemlock Looper (1), 13 and 19 August 1927, A.A. Wilks to Ontario Forestry Branch; see also 9 August 1927, E.J. Zavitz to A.A. Wilks; and 4 August 1927, G. Freemantle to Minister of Lands and Forests.

12 AO, RG1–256, 1, Hemlock Looper (1), 4 August 1927, G. Freemantle to Minister of Lands and Forests; see also 6 August 1927, W. Finlayson to G. Freemantle.

13 AO, RG1–256, 1, Hemlock Looper (1), 21 August 1927, F.J. Ames to Ontario Forestry Branch.

14 AO, RG1–256, 1, Hemlock Looper (1), 6 August 1927, W.H. Finlayson to E.J. Zavitz and 19 August 1927, E.J. Zavitz to P. McEwen.

15 AO, RG1–256, 1, Hemlock Looper (1), 24 and 29 August 1927, P. McEwen to Deputy Minister of Forestry, with the citations from the first document; AO, RG1–256, 1, Hemlock Looper (1), ca. 20 August 1927, J.J. de Gryse to A.H. Richardson.

16 AO, RG1–256, 1, Hemlock Looper (1), 29 August 1927, P. McEwen to E.J. Zavitz. The sawfly had practically wiped out Ontario's tamarack around the turn of the twentieth century.

17 AO, RG1–256, 1, Hemlock Looper (1), ca. 20 August 1927, J.J. de Gryse to A.H. Richardson.

18 AO, RG1–256, 1, Hemlock Looper (1), 29 August 1927, P. McEwen to F.J. Zavitz.

19 AO, RG1–256, 1, Hemlock Looper (1), 19 August 1927, F.C. Gratwick to Watson Jack Pump Company; 27 August 1927, Watson Jack Company to F.C. Gratwick; and 27 August 1927, F.C. Gratwick to J.J. de Gryse, from which the citation is taken.

20 AO, RG1–256, 1, Hemlock Looper (1), 30 August 1927, E.J. Zavitz to F.J. Ames, from which the first citation is taken; 30 August 1927, E.J, Zavitz to G. Freemantle, from which the second citation is taken; 31 August 1927, J.J. de Gryse to E.J. Zavitz; and 8 October 1927, E.J. Zavitz to H. Balm.

21 Canada Department of Agriculture, Entomological Branch, *The Canadian Insect Pest Review* 5, no. 4 (31 August 1927): 42.

22 AO, RG1–256, Box 1, Hemlock Looper (1), 4 October 1927, H. Balm to Forestry Department; see also 1 September 1927, W. Morris to Department of Agriculture; 15 September 1927, F.B. Robbins to W.H. Finlayson [telegram]; 28 September 1927, H. Balm to Department of Forestry; 2 February 1928, G.T. Clarkson to J.M. Swaine; 3 February 1928, W.R. Winter to J.M. Swaine; 21 February 1928, St. Clair Balfour to J.J. de Gryse; 29 February 1928, J. MacNeill to J.J. de Gryse; 29 February 1928, W.A. Campbell to J.J. de Gryse; and 7 March 1928, J.D. Remington to W.H. Finlayson.

23 AO, RG1–256, Box 1, Hemlock Looper (1), 2 February 1928, D.C. Mason to J.M. Swaine.

24 Ibid.

25 AO, RG1–256, Box 1, Hemlock Looper (1), 3 February 1928, E.R.C. Clarkson to J.M. Swaine, from which the first citation is taken; 20 February 1928, T.S. Welsman to E.H. Finlayson, from which the second citation is taken; and 13 March 1928, J. Poad to W.H. Finlayson, from which the third citation is taken.

26 AO, RG1–256, Box 1, Hemlock Looper (1), 16 February 1928, D.C. Mason to J.J. de Gryse.

27 AO, RG1–256, Box 1, Hemlock Looper (1), 14 April 1928, J.M. Swaine to E.J. Zavitz. See also LAC, RG17, 3131, 69 – Department of Militia and Defence, 14 April 1928, A. Gibson to J.H. Grisdale; AO, RG1–256, Box 1, Hemlock Looper (1), 1 February 1928, P. McEwen to J.J. de Gryse and 23 February 1928, J.M. Swaine to E.J. Zavitz.

28 Canada Department of Agriculture, Entomological Branch, *The Canadian Insect Pest Review* 6, no. 2 (25 June 1928): 17; and 6, no. 5 (5 October 1928): 46; *Forestry Chronicle* 5, no. 2 (1929):

35–41; AO, RG1–256, 1, Hemlock Looper (2), Project No. 2 – The Hemlock Looper Section A and Section D, enclosed in 16 July 1938, J.J. de Gryse to C.R. Mills; AO, RG-116, Box 3, Dusting Operations 1928, Spruce Budworm, 6 June 1928, J.M. Swaine to E.J. Zavitz.

29 Boyer, *Early Days in Muskoka*, 159.

30 AO, RG1–256, Box 1, Hemlock Looper (1), 7 March 1928, J.D. Remington to W.H. Finlayson; see also 13 March 1928, J. Poag to W.H. Finlayson (letters such as these delivered practically verbatim messages to Ontario's Minister of Lands and Forests); University of Waterloo Archives (UWA), GA100, Series 1, File 22 – Annual Meeting 1928, Muskoka Lakes Association, Minutes of the Annual Meeting held at the Muskoka Lakes Golf and Country Club, Lake Rosseau, Ontario, at 10 a.m. 13 August 1928.

31 UWA, GA100, Series 6, *Muskoka Lakes Association Yearbook, 1929*, from which the citation is taken; *Toronto Globe*, March 22, 1928; AO, RG1–256, Box 1, Hemlock Looper (1), 12 May 1928, E.J. Zavitz to C. Campbell (with attachment) and 23 June 1928, E.J. Zavitz to F.C. Gratwick.

32 The OPAS had been established in 1924 after Ontario decided to create its own air service instead of hiring contractors to provide flying services; it was focused on combatting forest fires. See West, *The Firebirds*.

33 Coad et al., *Dusting Cotton from Airplanes*; LAC, RG24-E-1-a, 4912, DND – Operations for Department of Agriculture, Vol. 2, correspondence from mid-1920s; AO, RG1–116, Box 3, Dusting Operations 1928, 19 March 1928, A.T. Cowley to W.R. Maxwell; and 20 March 1928, W.R. Maxwell to A.T. Cowley; AO, RG1–116, Box 3, Dusting Operations 1928, Memoranda – Deputy Minister, 20 March 1928, W.R. Maxwell to E.J. Zavitz.

34 LAC, RG24-1-a, 4912, DND – Operations for Department of Agriculture, Vol. 3, "Report on Use of Keystone Puffer for Dusting the Forest," 28 March 1928, C.L. Bath to Director, Civil Government Air Operations.

35 LAC, RG39, 341, 47278, 10 April 1928, J.L. Gordon to E.H. Finlayson.

36 West, *The Firebirds*, 147.

37 LAC, RG39, 341, 47278, 12 April 1928, E.H. Finlayson to W.H. Finlayson; AO, RG-116, Box 3, Dusting Operations 1928, 17 April 1928, W.H. Finlayson to W.R. Maxwell; AO, RG-116, Box 3, Dusting Operations 1928, Spruce Budworm, 6 June 1928, J.M. Swaine to E.J. Zavitz; AO, RG-116, Box 3, Dusting Operations 1928, Memoranda – Deputy Minister, 20 March and 10 April 1928, W.R. Maxwell to E.J. Zavitz; AO, RG-116, Box 3, Dusting Operations 1928, "Prices of American Aircraft"; AO, RG1–256, Box 1, Spruce Budworm, 28 April 1928, E.J. Zavitz to J.M. Swaine; West, *The Firebirds*, chap. 11.

38 AO, RG1–256, Box 1, Hemlock Looper (1), 23 June [telegram] and 3 July 1928, J.J. de Gryse to E.J. Zavitz.

39 AO, RG1–256, 1, Hemlock Looper (1), 11 June 1928, F.C. Gratwick to Department of Forestry and 23 June 1928, E.J. Zavitz to F.C. Gratwick.

40 AO, RG1–256, Box 1, Hemlock Looper (1), 13 July 1928, F.C. Gratwick to E.J. Zavitz; see also 16 June 1928, P. McEwen to E.J. Zavitz; 14 July 1928, A.W. Sawyer to E.J. Zavitz; and 16 July 1928, G.H. Chisholm to Mr. Vavitz [*sic*].

41 AO, RG1–256, Box 1, Hemlock Looper (1), 12 July 1928, F.J. Capon to E.J. Zavitz.

42 AO, RG1–256, Box 1, Hemlock Looper (1), 12 July 1928, E.J. Zavitz to F.J. Capon [telegram]; see also 1 February 1928, P. McEwen to J.J. de Gryse; 11 May 1928, J.M. Swaine to E.J. Zavitz; 18 May 1928, E.J. Zavitz to J.M. Swaine; 9 June 1928, Cowan Chemical Company to E.J. Zavitz; 19 June 1928, J.J. de Gryse to E.J. Zavitz; 20 June 1928, E.J. Zavitz to Cowan Chemical Company; 9 July 1928, P. McEwen to E.J. Zavitz; and 12 July 1928, F.J. Capon to E.J. Zavitz [telegram].

43 AO, RG1–256, Box 1, Hemlock Looper (1), 12 July 1928, E.J. Zavitz to W.R. Maxwell [telegram] and 12 July 1928, W.R. Maxwell to E.J. Zavitz [telegram].

44 AO, RG1–256, Box 1, Hemlock Looper (1), 16 July 1928, G.H. Chisholm to Mr. Vavitz [*sic*].

45 AO, RG1–256, Box 1, Hemlock Looper (1), 13 July 1928, E.J. Zavitz to Officer in Charge, RCAF [telegram].

46 AO, RG1–256, Box 1, Hemlock Looper (1), 14 July 1928, Civil Operations, RCAF, to E.J. Zavitz [telegram].

47 AO, RG1–116, Box 3, Dusting Operations 1928, 13 July 1928, Deputy Minister of Forestry to W.R. Maxwell [telegram], from which the first citation is taken; 13 July 1928 [2], W.R. Maxwell to E.J. Zavitz [telegram], from which the second citation is taken; and 13 July 1928 [2], W.R. Maxwell to R.E. Nicoll [telegram], from which the final citation is taken; AO, RG1–256, Box 1, Hemlock Looper (1), 13 July 1928, W.R. Maxwell to E.J Zavitz [telegram].

48 AO, RG1–116, Box 3, Dusting Operations 1928, ca. July 1928, RCAF – Issue Order for Equipment [re: loaning Keystone Puffer to Ontario Provincial Air Service].

49 AO, RG1–116, Box 3, Dusting Operations 1928, 16 July 1928, G.R. Hicks to Director, Provincial Air Service [telegram].

50 AO, RG1–116, Box 3, Dusting Operations 1928, 18 July 1928, W.R. Maxwell to G.R. Hicks [telegram].

51 AO, RG1–116, Box 3, Dusting Operations 1928, 18 July 1928, G.R. Hicks to W.R. Maxwell.

52 AO, RG1–116, Box 3, Dusting Operations 1928, 26 July 1928, G.R. Hicks to W.R. Maxwell.

53 AO, RG1–256, Box 1, Hemlock Looper (1), "Summary Report on Airplane Dusting Operations, Footes Bay, Ontario, 1928," enclosed in 19 October 1928, J.J. de Gryse to C.R. Mills.

54 AO, RG1–116, Box 3, Dusting Operations 1928, ca. 19 July 1928, G.R. Hicks to W.R. Maxwell.

55 West, *The Firebird*, 149.

56 AO, RG1–116, Box 3, Dusting Operations 1928, 26 July 1928, G.R. Hicks to W.R. Maxwell.

57 AO, RG1–256, Box 1, Hemlock Looper (1), "Summary Report on Airplane Dusting Operations, Footes Bay, Ontario, 1928," enclosed in 19 October 1928, J.J. de Gryse to C.R. Mills; AO, RG1–116, Box 3, Dusting Operations 1928, 19 July 1928, W.R. Maxwell to G.R. Hicks [telegram].

58 Canada Department of National Defence, *Report on Civil Aviation and Civil Government Air Operations for the Year 1928* (Ottawa: F.A. Acland, Printer to the King's Most Excellent Majesty, 1929).

59 AO, RG1–256, Box 1, Hemlock Looper (1), 21 July 1928, J.J. de Gryse to C.R. Mills [telegram].

60 AO, RG1–256, Box 1, Hemlock Looper (1), 22 Ju[ly] 1928, J.J. de Gryse to C.R. Mills.

61 AO, RG1–256, Box 1, Hemlock Looper (2), 19 October 1928, "Summary Report on Airplane Dusting Operations, Footes Bay, Ontario, 1928," J.J. de Gryse.

62 AO, RG1–116, 3, Dusting Operations 1928, 26 July 1928, G.R. Hicks to W.R. Maxwell.

63 AO, RG1–256, 1, Hemlock Looper (1), 3 August 1928, J. Penrose to Ontario Department of Forests and Mines; 14 August 1928, W.E. Taylor to P.W. Hodgetts; and 25 August 1928, E.A. Apple to Department of Forests and Mines, from which the citations are taken.

64 AO, RG1–256, 1, Hemlock Looper (1), 14 August 1928, O.W. Clapperton to Department of Biology, Ontario, from which the citations are taken; see also AO, RG1–256, 1, Hemlock Looper (2), 1 October 1928, J.J. Marshall to Minister of Lands and Forests; 20 April 1929, D.C. Donaldson to C.R. Mills; 30 April 1929, C.R. Mills to D.C. Donaldson; and 30 April 1929, C.R. Mills to W.A.P. Wood.

65 AO, RG1–256, 1, Hemlock Looper (2), 20 August 1928, E.B. Whitcomb to Provincial Forester.

66 AO, RG1–256, 1, Hemlock Looper (1), 20 November 1928, E.B. Whitcomb to G.W. Ecclestone, enclosed in 28 November 1928, G.W. Ecclestone to W.H. Finlayson; see also 5 December 1928, C.R. Mills to Whitcomb.

67 AO, RG1–256, 1, Hemlock Looper (1), 3 August 1928, J. Penrose to C.R. Mills.

68 AO, RG1–256, 1, Hemlock Looper (1), 11 August 1928, C.R. Mills to J. Penrose; see also 20 August 1928, C.R. Mills to O.W. Clapperton; 22 August 1928, C.R. Mills to E.B. Whitcomb; 23

August 1928, C.R. Mills to T.S. Cullen; 25 August 1928, C.R. Mills to J.B. Jarvis; and 27 August 1928, C.R. Mills to E.A. Apple and separate letter to W.E. Taylor.

69 AO, RG1–256, 3, Dusting Operations 1928, 30 August 1928, W.R. Maxwell to C.R. Mills.

70 AO, RG1–256, 1, Hemlock Looper (1), 25 July, 6, 7 and 22 September 1928, C.R. Mills to J.J. de Gryse; 30 July and 4 September 1928, J.J. de Gryse to C.R. Mills; 30 July 1928, C.R. Mills to P. McEwen; 9 August 1928, C.R. Mills to B.R. Coad; 13 August 1928, R.C. Gaines to C.R. Mills; 25 August 1928, C.R. Mills to W.R. Maxwell; 30 August and 21 September 1928, W.R. Maxwell to C.R. Mills; 4 September 1928, J.J. de Gryse to C.R. Mills; 15 September 1928, P. McEwen to Deputy Minister of Forestry; 24–6 September 1928, telegrams between Ottawa, Toronto, and Sault Ste. Marie regarding the dusting project; and 29 September and 1 October 1928, F.A. MacDougall to Deputy Minister of Forestry; AO, RG1–256, 3, Dusting Operations 1928, 30 August and 14 September 1928, W.R. Maxwell to J. Hyde; 6 September 1928, G.B. Holmes to W.R. Maxwell; 7 September 1928, W.R. Maxwell to Provincial Air Service – Sault Ste. Marie (PASSSM); 7 September 1928, PASSSM to W.R. Maxwell; 8 and 24 September 1928, C.R. Mills to W.R. Maxwell; 8 and 15 September 1928, J. Hyde to W.R. Maxwell; 10 September 1928, C.R. Mills to P. McEwen; 11 September and 1 October 1928, W.H. Ptolemy to W.R. Maxwell; 21 September and 4 October 1928, W.R. Maxwell to C.R. Mills; 24 September 1928, PASSSM to C.R. Mills; 28 September 1928, W.R. Maxwell to G.R. Hicks; 28 September and 1 October 1928, W.H. Ptolemy to W.R. Maxwell; and 1 October 1928, G.R. Hicks to W.R. Maxwell; AO, RG1–256, 1, Hemlock Looper (2), 1 October 1928, W.R. Maxwell to C.R. Mills; 11 October 1928, C.R. Mills to J.J. de Gryse; and 18 March 1929, E.J. Zavitz to W.R. Maxwell.

71 AO, RG1–256, 1, Hemlock Looper (1), 3 August 1928, J. Penrose to Ontario Department of Forests and Mines [sic]; 11 August 1928, C.R. Mills to J. Penrose; 11 August 1928, C.R. Mills to P. McEwen; 14 August 1928, W.E. Taylor to P.W. Hodgetts; 21 and 25 August 1928, T.S. Cullen [President – Ahmic Lake Cottagers Association] to W. Finlayson; 23 August 1928, J.B. Jarvis to Department of Lands and Forests; 11 September 1928, P. McEwen to Deputy Minister of Forestry; and 18 September 1928, Acting Deputy Minister of Forestry to P. McEwen; AO, RG1–256, 1, Hemlock Looper (2), 14 November 1928, C.R. Mills to F.A. MacDougall, with notation; 9 January 1929, T. Price to Forestry Department; 1 March 1929, W.D. Steele to Ontario Department of Lands and Forests; 4 March 1929, E.J. Zavitz to W.D. Steele; 2 April 1929, J. Penrose to Minister of Forestry; 9 April 1929, C.R. Mills to J. Penrose; 8 April 1929, C. Campbell to Deputy Minister of Forestry; and 16 April 1929, C.R. Mills to C. Campbell.

72 AO, RG1–256, 1, Hemlock Looper (2), 21 May 1929, W.C. Holt to W.H. Finlayson; 27 May 1929, C.R. Mills to W.C. Holt; 1 June 1929, Acknowledgement of Order from Deloro Chemical Company Limited; and 22 April 1929, P. McEwen to Deputy Minister; "Timber Insects Will Dine on $6,000 Meal of Poison," *Globe*, February 27, 1929.

73 Concerns were also expressed about the dangers of using the DH61 for dusting work because it was too cumbersome and its wingspan too broad but nothing came of them because this was the only time that this aircraft was used for this type of work in Canada: West, *The Firebird*, 150.

74 It does not appear that the government dusted the looper outbreak in the Ahmic Lake area even though it had been raised at a meeting of the local cottagers' association in August 1928: Ahmic Lake Cottage Owners' Association – Minutes of Annual Meeting held 15 August 1928 at Dr. Thomas S. Cullen's camp. I am indebted to Douglas W. Reid (Vice-president – Membership for the Ahmic Lake Cottagers Association) for his assistance with my research.

75 Ontario Department of Lands and Forests, *Report of the Minister of Lands and Forests of the Province of Ontario For the Year Ending October 31st, 1929* (Toronto: Herbert H. Ball, Printer to the King's Most Excellent Majesty, 1930), 151; AO, RG1–256, 1, Hemlock Looper (2), 16 July 1938, J.J. de Gryse to C.R. Mills, to which is attached "Project No. 1 – The Hemlock Looper

Section A and Section D" (this paragraph and the one that follows are based largely on this source); AO, RG1–256, 1, Hemlock Looper 193[0], 3 and 17 November 1930, P. McEwen to Deputy Minister of Forestry. The sudden termination of the effort in Georgian Bay created a surplus of roughly 5,000 pounds of calcium arsenate, which the Ontario government sold to farmers in southern Ontario.

76 AO, RG1–256, 1, Hemlock Looper (2), 16 July 1938, J.J. de Gryse to C.R. Mills, to which is attached "Project No. 1 – The Hemlock Looper Section A and Section D"; and 16 July 1929, J.J. de Gryse to E.J. Zavitz, from which the citation is taken; J.J. de Gryse and K. Schedl, "An Account of the Eastern Hemlock Looper, Eloppia fiscellaria Gn., on Hemlock, with Notes on Allied Species," *Scientific Agriculture* 14, no. 10 (1934): 523–39.

77 Gibson, "The Canadian Entomological Service," 1460.

5 "You cannot control an infestation such as this with toys"

1 Drushka, *HR*, 68–82. British Columbia also developed a significant pulp and paper industry, but it paled in comparison to the province's lumber industry. The opposite situation prevailed in eastern Canada, where the pulp and paper sector soon dwarfed the lumber industry, which had previously been the country's dominant forest enterprise.

2 Whitford and Craig, *The Forest Resources of British Columbia*, 7; Roach, "Stewards of the People's Wealth"; Drushka, *HR*; Gillis and Roach, *Lost Initiatives*, 145–9.

3 British Columbia Archives (BCA), GR-1441, 04214 #1, 13 March 1913, C.G. Hewitt to W.R. Ross; see also 22 March 1913, J.M. Swaine to H.R. MacMillan.

4 BCA, GR-1441, 04214 #1, 14 March 1913, H.R. MacMillan to J.M. Swaine. The more that foresters and entomologists alike investigated BC's insect infestations, the more they realized that these outbreaks had always been an inherent part of BC's forests. George R. Hopping does a great job of tracing the history of hemlock looper infestations in the province in his MSc thesis, "The Western Hemlock Looper," 8–10.

5 British Columbia Forest Service Archives (BCFSA), File 048811 MFP Insects, 6 May 1913, H.R. MacMillan to G.D. McKay; and 8 May 1913, G.D. McKay to H.R. MacMillan; BCA, GR-1441, 04214 #1, 25 April and 13 and 16 June 1913, H.R. MacMillan to J.M. Swaine; 27 April 1913, J.M. Swaine to H.R. MacMillan; April–May correspondence from H.R. MacMillan to BC's District Foresters; 31 May 1913, C.G. Hewitt to H.R. MacMillan; and 16 June 1913, H.R. MacMillan to *Pulp and Paper Magazine of Canada*, *Western Lumberman*, *Canada Lumberman*, *Canadian Forestry Journal*, North Pacific Lumber Company, Forest Mills of British Columbia, Brooks, Scanlon & O'Brien Logging Company, and Victoria Lumber Company.

6 BCA, GR-1441, 04214 #1, 25 April 1913, H.R. MacMillan to C.G. Hewitt.

7 Swaine, *Forest Insect Conditions in British Columbia*; BCA, GR-1441, 04214 #1, 10 April 1913, Assistant District Foresters, Portland, Oregon to Forest Supervisor; [?] October 1913, article in Oregon newspaper regarding the BCFB aggressively burning beetle infested trees and slash. A minor amount of slash burning occurred that season, and thereafter an increasing volume of work was undertaken in an effort to control bark beetle infestations; Rajala, "The Vernon Laboratory," 177–8.

8 City of Vancouver Archives (CVA), PDS-12–1, 609, Annual Reports of Vancouver Board of Park Commissioners, 1913–1916, "Annual Report for 1913." Kheraj, *Inventing Stanley Park*, chaps. 1–3, thoroughly traces Stanley Park's history and demonstrates how myriad human hands dramatically manipulated it. Understandably, however, Kheraj's focus was the park, and not the chemical spraying and dusting projects that occurred there, so he did not contextualize the latter events within the broader framework of developments in Canadian forest entomology.

Furthermore, he relied largely upon sources located in Vancouver, and thus his story minimizes the influence exerted on events in the park by officials at both the provincial and dominion levels.

9 Spirn, "Constructing Nature," 91; Fauvreau, "Understanding Natural Constructivism," passim.

10 CVA, VPK-S97, MSS50-E-4, Park Board – Clippings, 1911–1925, clippings from 1912, with the citation coming from 13 January 1912, *Saturday Sunset*, Editorial. Kheraj, *Inventing Stanley Park*, chap. 2, deals with this subject and the inaugural attempt to knock back an insect infestation in Stanley Park in 1913–14.

11 BCA, GR-1441, 04221, 16 October 1913, G.A. McGuire to W.R. Ross; 22 and 24 October 1913, W.R. Ross to G.A. McGuire.

12 BCA, GR-1441, 04221, 22 October 1913, W.R. Ross to H.R. MacMillan; BCFSA, File 048811 – MFP Insects, 22 December 1913, W.H. Baggs to C.H. Edgecombe; CVA, VPK, S81, MSS48-C-5, Park Board Correspondence – Department of Agriculture – Stanley Park, December 1913–May 1914, 22 December 1913, Secretary to C.H. Edgecombe.

13 LAC, MG27-III-B-9, Finding Aid 141, H.H. Stevens; CVA, VPK, S81, MSS48-C-5, Park Board Correspondence – Forestry – Stanley Park Oct. 1913–Sept. 1919, 15 October 1913, W.S. Rawlings to H.H. Stevens and 24 October 1913, H.H. Stevens to W.R. Owen.

14 Gray, "The Government Timber Business"; BCA, GR-1441, 04221, November 3, 1913, *Vancouver Daily Province*; December 13, 1913, *The World*; February 2 and 11, 1914, *Vancouver Daily Province*; November 28, 1913, *Vancouver Sun*; and September 18, 1913, *Vancouver Daily Province*; also see September 19, 1913, *Vancouver Sun*; CVA, VPK, S76, MSS48-A-3, Park Board Minutes, May 1913–April 1917, The Minutes of the Regular Meeting of the Board of Park Commissioners Held in the Park Board Offices [hereafter Minutes] on 8 October 1913 at 3pm; Minutes, 12 November 1913; Minutes, 29 December 1913; Minutes, 11 February 1914; and Minutes, 25 February 1914; Kheraj, *Inventing Stanley Park*, chap. 2.

15 CVA, VPK, S81, MSS48-C-5, Park Board Correspondence – Forestry – Stanley Park Oct. 1913–Sept. 1919, ca. January 1914, "Preliminary Report on Insect Conditions in Stanley Park, Vancouver, British Columbia."

16 CVA, VPK, S81, MSS48-C-5, Park Board Correspondence – Forestry – Stanley Park Oct. 1913–Sept. 1919, 31 January 1914, H.H. Stevens to W.R. Owen, enclosing January 1914, Memorandum by Dominion Botanist Regarding Conditions of Health of Trees, Stanley Park, Vancouver, B.C. and January 1914, Memorandum on the Damage Due to Insects in Stanley Park, Vancouver, British Columbia, C.G. Hewitt; BCA, GR-1441, 04221, 2 January 1914, C.H. Edgecombe.

17 The plan to clear the brush from Stanley Park had already unleashed a major protest from local citizens concerned about keeping it "natural": BCA, GR-1441, 04214 #1, 14 January 1914, J.M. Swaine to H.R. MacMillan.

18 BCA, GR-1441, 04214 #1, 14 January 1914, J.M. Swaine to H.R. MacMillan.

19 Ibid.

20 BCA, GR1441, 04214 #1, 18 April 1914, J.M. Swaine to H.R. MacMillan, from which the citation is taken; see also 29 April 1914, H.R. MacMillan to J.M. Swaine and 20 May 1914, C.G. Hewitt to H.R. MacMillan; BCFSA, File 048811 – MFP Insects, 24 February 1914, H.R. MacMillan to J.M. Swaine.

21 CVA, VPK, S76, MSS48-A-3, Park Board Minutes, May 1913–April 1917, Minutes, 14 and 28 April 1915; Minutes, 26 May 1915; Minutes, 9 June 1915; Minutes, 14 July 1915; and Minutes, 27 October 1915; BCA, GR-1441, 04214 #1, ca. December 1914, Forest Investigations in British Columbia, Entomological Branch; BCA, GR-1441, 04214 #1, documents from 1916–1917 concerning R.N. Chrystal's work in Stanley Park.

22 The Park Board had a few crucial questions about the project, namely which administrative body had the authority to approve it and who was going to foot the bill for it. Ultimately it was determined that the dominion Department of Militia and Defence had jurisdiction over this work, and after much delay, it agreed to pay for it: December 13, 1913, *The World*; February 2 and 11 and June 24, 1914, *Vancouver Daily Province*; June 11 and 25, 1914, *Vancouver Sun*; CVA, VPK, S76, MSS48-A-3, Park Board Minutes May 1913–April 1917, Minutes of meetings held 11 and 25 February, 10 June, 22 July, 29 September, and 28 October 1914; and 24 February 1915; CVA, VPK, S81, MSS48-C-5, Park Board Correspondence – Entomological Department, 1913–1915, all correspondence but especially 9 May 1914, W.S. Rawlings to J.M. Swaine; 22 May 1914, W.R. Owens to A.E. Lees; 30 June and 30 September 1914, W.S. Rawlings to J.M. Swaine; 1 July 1914, J.M. Swaine to W.S. Rawlings; 2 October 1914, W.S. Rawlings to S. Hughes; and 19 October 1914, C.F. Winter to W.H. Baggs.

23 CVA, VPK, S81, MSS48-C-5, Park Board Correspondence – Entomological Department, 1913–1915, 11 December 1913, J.M. Swaine to W.R. Ross. Although this type of work was novel in British Columbia's forests, the province had long been using toxins to kill unwanted pests in agriculture. This was particularly true in BC's fruit-growing industry, whose size and economic importance had propelled it to the leading edge of spraying work in Canada by this time: BCA, GR-1190, 1, Plant Pathologists Records, 1915–1933, all documents from Files 1–3; Thistle, "Accommodating Cattle"; CVA, VPK, S81, MSS48-C-5, Park Board Correspondence – Entomological Department, 1913–1915, March Records of Experiments in Spraying Stanley Park, Vancouver, British Columbia; 7 February and 19 March [2] 1914, A.W. Latterfield to Board of Park Commissioners; 16 and 23 February 1914, W.S. Rawlings to J.M. Swaine; 18 February 1914, C.G. Hewitt [2]; 2 March 1914, C.G. Hewitt to R.C. Treherne; 27 March 1914, M.A. Grainger to W.S. Rawlings; and 4 April 1914, W.S. Rawlings to M.A. Grainger; CVA, VPK, S64, MSS48-A-3, Park Board Minutes May 1913–April 1917, Minutes, 26 March 1914 and Minutes, 8 April 1914; BCA, GR-1441, 04221, 27 February and 6 March 1914, H.R. MacMillan to Board of Park Commission; 11 March 1914, M.A. Grainger to J.B. Mitchell; 30 March 1914, J.B. Mitchell to M.A. Grainger, enclosing 25 March 1914, "Report on Silvicultural Conditions in Stanley Park"; 17 March 1915, W.H. Baggs to H.R. MacMillan; March 20, 1914, *Vancouver Daily Province*; and March 24, 1913, *Vancouver Sun*.

24 CVA, VPK, S81, MSS48-C-5, Park Board Correspondence – Entomological Department, 1913–1915, 19 March 1914, J.M. Swaine to W.S. Rawlings and 4 April 1914, W.S. Rawlings to J.M. Swaine.

25 CVA, VPK, S1, PDS-12-1, 609, Annual Reports of Vancouver Board of Park Commissioners, 1913–1916, "Annual Report for 1913" and "Annual Report for 1916"; CVA, VPK, S81, MSS48-C-5, Park Board Correspondence – Entomological Department, 1913–1915, 23 September 1914, J.M. Swaine to W.R. Owen; 27 October 1914, "Treatment of the Beetle Attacked Spruce in Stanley Park," R.N. Chrystal; CVA, VPK, S76, MSS48-A-3, Park Board Minutes, May 1913–April 1917, 27 October 1915, Monthly Report by W.S. Rawlings, Park Superintendent; BCA, GR1441, 04214 #1, 8 August and 6 October 1914, J.M. Swaine to H.R. MacMillan; ca. October 1914, Forest Investigations in British Columbia, Entomological Branch; and 18 March 1915, J.M. Swaine to H.R. MacMillan.

26 February 17, 1914 and February 9, 1915, *Vancouver Sun*; February 13, 1915, *Vancouver Daily Province*.

27 CVA, VPK, S81, MSS48-C-5, Park Board Correspondence – Entomological Department, 1913–1915, 3 June 1914, W.S. Duncan to Park Board, from which the citation is taken; and 11 June 1914, W.S. Rawlings to W.S. Duncan.

28 CVA, VPK, S81, MSS48-C-5, Park Board Correspondence – Entomological Department, 1913–1915, 30 March 1915, W.S. Rawlings to T. Cunningham.

29 CVA, VPK, S1, PDS-12–1, 609, Annual Reports of Vancouver Board of Park Commissioners, 1913–1916, "Annual Report for 1915," from which the citation is taken; see also "Annual Report for 1916."

30 BCFSA, File 048811 – MFP Insects, all correspondence from 1915 to 1918; BCA, GR-1441, 04214 #1, all correspondence between May 1916 and October 1919; BCA, GR-1441, 04216, 04217 and 04220, all documents; Swaine, *Canadian Bark-Beetles: Part I* and *Part III*; Richmond, "A History of Forest Entomology in British Columbia."

31 BCA, GR-1441, 04214 #1, 9 April 1920, J.M. Swaine to M.A. Grainger.

32 Rajala, "The Vernon Laboratory," 183.

33 Entomological Society of British Columbia, *Proceedings for 1941*, "Ralph Hopping (1868–1943)," 3–4; BCA, GR-1441, 04214 #1, 17 November 1919, J.M. Swaine to M.A. Grainger and 29 November 1919, M.A. Grainger to R.D. Prettie; Rajala, "The Vernon Laboratory," 178–83. Hopping's career is the subject of M. Kuhlberg, "'one h--- of a fellow': The Remarkable Career of Ralph Hopping, Pioneering Forest Entomologist, 1868–1941," unpublished manuscript.

34 Drushka, *HR*, 143.

35 BCFSA, File 048811 – MFP Insects, correspondence from 1920s; BCA, GR-1441, 04214 #1 and #2, correspondence from 1920s; BCA, GR-1441, 04216, all correspondence, which is summarized in M. Kuhlberg, "Be Careful What You Wish For: Government-Industry Relations in a Beetle Infested Forest in British Columbia's Interior, 1916–1925," unpublished manuscript; BCA, GR-1230, 1, 1/1, "British Columbia Forest Service, An Outline of the Work of the Research Division."

36 For example, see Hopping, *The Control of Bark-Beetle Outbreaks in British Columbia*; California Academy of Sciences (CAS), Ralph Hopping Papers (RHP), 325, W. Downes, 13 July 1926, W. Downes to R. Hopping and 27 July 1926, W. Downes to R. Hopping; Rajala, "The Vernon Laboratory," 178–9.

37 *Proceedings of the 22nd Annual Meeting of the Entomological Society of Alberta*, 39–40.

38 Hidy et al., *Timber and Men*; *Canada Lumberman and Woodworker*, January 1, 1912, 57; "Edward Young, Timber Broker, Dies at Age 71," *Wisconsin State Journal*, September 27, 1948.

39 LAC, RG15, D-V-1, Vol. 1262, 505018 – Brittingham & Young Co Ltd – Timber Berths 469, 389, and 510, all documents; BCA, GR-3100, B04700 and B04705, Timber Licence Files for STL8 and STL9, all documents.

40 LAC, RG13-a-2, Vol. 1921, 1909–475 and 1912–752 – Brittingham & Young, Application for Booming Privileges on Indian River, 23 March 1909, Secretary, Department of Public Works to E.L. Newcombe and 26 March 1909, E.L. Newcombe to Secretary, Department of Public Works; LAC, RG13-a-2, Vol. 168, 1911–1903 – Draft Lease to Brittingham & Young for Logging Purposes 1911, 25 June 1910, L. Pereira to T.R.E. McInnes and 29 September 1911, T.R.E. McInnes to L. Pereira; LAC, RG13-a-2, Vol. 165, 1910–1316 and 1911–239 – Brittingham & Young – Use of 3 Streams for Logging, 25 October 1910, Assistant Secretary, Department of Interior to E.L. Newcombe (with attachments); BCA, GR-0436, B14628, Carton 7858 – Brittingham & Young, 29 July 1911, Thompson and Burgess to Department of Interior.

41 BCA, GR-0385, 11, Files 81 and 91, all documents.

42 Humphreys and Wong, *The History of Wigwam Inn*, chaps. 1–4; LAC, RG13-a-2, 300, Brittingham & Young, 1925–1360, all documents.

43 "Wigwam Inn Ends Big Trip," *Vancouver Sun*, June 25, 1926.

44 "Wigwam Inn Has Atmosphere," *Vancouver Sun*, August 12, 1927.

45 Canada Department of Agriculture, Entomological Branch, *The Canadian Insect Pest Review* 6, no. 4 (25 August 1928): 39, from which the citation is taken; see also Canada Department of

Agriculture, Entomological Branch, *The Canadian Insect Pest Review* 7, no. 3 (15 July 1929): 33; CAS, RHP, 326, E. Hearle, 5 November 1928, E. Hearle to R. Hopping, enclosing "Notes on the Outstanding Forest Insects in British Columbia for the Season 1928"; BCFSA, File 048811 – MFP Insects, 6 July 1929, "Report by Mr. Hopping – Dominion Entomologist, Summary of Hemlock Looper Infestation, Indian River, British Columbia."

46 PFCA, Project Dusting – Wigwam Inn 1929, 21 September 1928, G.R. Hopping to W.J. Chamberlin.

47 PFCA, Project Dusting – Wigwam Inn 1929, 27 May 1929, G.R. Hopping to R. Hopping, from which the first two citations are taken; 31 May 1929, R. Hopping to G.R. Hopping; 28 May 1929, G.R. Hopping to Brittingham & Young, from which the third and fourth citations are taken; and 3 June 1929, E.J. Young to G.R. Hopping, from which the fifth citation is taken.

48 Hopping, "The Western Hemlock Looper," 36; PFCA, Project Dusting – Wigwam Inn 1929, 3 June 1929, E.J. Young to G.R. Hopping.

49 PFCA, Project Dusting – Wigwam Inn 1929, ca. 3 June 1929, G.R. Hopping to E.J. Young; Hopping delivered practically the same message roughly a week later: ca. 12 June 1929, G.R. Hopping to E.J. Young.

50 PFCA, Project Dusting – Wigwam Inn 1929, 10 June 1929, G.R. Hopping to District Forester, Vancouver District.

51 PFCA, Project Dusting – Wigwam Inn 1929, 10 June 1929, E.J. Young to G.R. Hopping; 26 June 1929, G.R. Hopping to E.J. Young, from which the first two citations are taken; 26 June 1929, J.R. Grant to G.R. Hopping; and 26 June 1929, G.R. Hopping to Grant, from which the remaining citations are taken.

52 PFCA, Project Dusting – Wigwam Inn 1929, 28 June 1929, G.R. Hopping to District Forester, Vancouver District.

53 PFCA, Project Dusting – Wigwam Inn 1929, 2 July 1929, A.E. Parlow to G.R. Hopping.

54 PFCA, Project Dusting – Wigwam Inn 1929, 4 July 1929, G.R. Hopping to A.E. Parlow. Parlow tried to defuse the situation in his reply to Hopping: 5 July 1929, A.E. Parlow to G.R. Hopping.

55 PFCA, Project Dusting – Wigwam Inn 1929, 8 July 1929, G.R. Hopping to J.M. Swaine, from which the citation is taken; see also 10 July 1929, J.R. Grant to G.R. Hopping; BCFSA, File 048811 – MFP Insects, 2 July 1929, G.R. Grant to P.Z. Caverhill; 5 July 1929, P.Z. Caverhill to R. Hopping; 5 July 1929, P.Z. Caverhill to G.R. Grant; Gray, "The Government Timber Business," passim.

56 PFCA, Project Dusting – Wigwam Inn 1929, 8 July 1929, G.R. Hopping to J.M. Swaine, from which the citations are taken; see also 10 July 1929, J.R. Grant to G.R. Hopping.

57 PFCA, Project Dusting – Wigwam Inn 1929, 10 June 1929, G.R. Hopping to Western Canada Airways; 10 June 1929, C.H. Daniels to G.R. Hopping; 12 and 18 June 1929, J.C. Barr to G.R. Hopping; 17 June 1929, C.C.B. Cave to G.R. Hopping; 18 June 1929, G.R. Hopping to R. Hopping; 22 June 1929, R. Hopping to G.R. Hopping; 27 June 1929, The British American Chemical Company Limited to G.R. Hopping; and 12 July 1929, Oliver Chemical Company to G.R. Hopping [2].

58 LAC, RG24–1-a, 4913, DND – Operations for Department of Agriculture, Vol. 4, 12 July 1929, Civil Air Operations to J.M. Swaine.

59 BC's Chief Forester, P.Z. Caverhill, aggressively pushed Hopping to arrange for the trimotor Ford to be used in BC, largely because it would relieve him of any obligation to obtain a plane for use in aerial dusting: BCFSA, File 048811 – MFP Insects, 5 July 1929, P.Z. Caverhill to R. Hopping; 5 July 1929, P.Z. Caverhill to J.R. Grant; and 6 July 1929, "Report by Mr. Hopping – Dominion Entomologist, Summary of Hemlock Looper Infestation, Indian River, British Columbia"; PFCA, Project Dusting – Wigwam Inn 1929, 13 July 1929, R. Hopping to G.R. Hopping; 14 June 1929, G.R. Hopping to Dominion Airways; 15 June 1929, L.A. Dobbin to G.R. Hopping; 18 June and

11 July 1929, D.R. MacLaren (Western Canada Airways) to G.R. Hopping; and 2 July 1929, G.R. Hopping to D.R. MacLaren; LAC, RG24–1-a, 4913, DND – Operations for Department of Agriculture, Vol. 4, 11 July 1929, J.M. Swaine to J.L. Gordon [telegram].

60 PFCA, Project Dusting – Wigwam Inn 1929, 18 July 1929, D.R. MacLaren to G.R. Hopping.

61 PFCA, Project Dusting – Wigwam Inn 1929, 18 July 1929, D.R. MacLaren to G.R. Hopping; 16 October 1929, W.H. Lyne to Campbell & Grill, enclosing 31 July 1929, Campbell & Grill to Entomological Branch; 1 August 1929, G.R. Hopping to A.E. Parlow; 17 October 1929, Campbell & Grill to Entomological Branch; and 18 October 1929, R. Hopping to G.R. Hopping.

62 BCFSA, File 048811 – MFP Insects, 11 July 1929, J.M. Swaine to P.Z. Caverhill; 12 July 1929, P.Z. Caverhill to J.M. Swaine [telegram]; PFCA, Project Dusting – Wigwam Inn 1929, 19 July 1929, G.R. Hopping to D.R. MacLaren, from which the citation is taken.

63 PFCA, Project Dusting – Wigwam Inn 1929, 24 July 1929, G.R. Hopping to A.E. Parlow; BCFSA, File 048811 – MFP Insects, 22 July 1929, A.E. Parlow to P.Z. Caverhill.

64 BCFSA, File 048811 – MFP Insects, 15 August 1929, A.E. Parlow to P.Z. Caverhill, enclosing 12 August 1929, A.E. Parlow to P.Z. Caverhill; LAC, RG39, 341, 47278, 7 August 1929, "Preliminary Report of Airplane Dusting at Wigwam Inn, Burrard Inlet, British Columbia," G.R. Hopping; Hopping, "The Western Hemlock Looper," passim.

65 PFCA, Project Dusting – Wigwam Inn 1929, 1 and 6 August 1929, G.R. Hopping to A.E. Parlow, with the citation taken from the second document; see also 9 August 1929, A.E. Parlow to G.R. Hopping; BCFSA, File 048811 – MFP Insects, 17 August 1929, E.B. Prowd to A.E. Parlow.

66 PFCA, Project Dusting – Wigwam Inn 1929, 7 August 1929, G.R. Hopping to Brittingham & Young; see also 8 August 1929, G.R. Hopping to J.R. Grant and 9 August 1929, J.R. Grant to G.R. Hopping.

67 PFCA, Project Dusting – Wigwam Inn 1929, 5 August 1929, R. Hopping to G.R. Hopping; and 8 August 1929, G.R. Hopping to R. Hopping, from which the citations are taken.

68 PFCA, Project Dusting – Wigwam Inn 1929, 7 August 1929, E. Walmsley to G. Hopping.

69 PFCA, Estimates [1929–1930] – 2012–08–01–019, "Estimates" [for 1929–1930].

70 Hopping openly scolded B&Y for its negligence in his letter to the company after the dusting project was completed: PFCA, Project Dusting – Wigwam Inn 1929, 7 August 1929, G.R. Hopping to Brittingham & Young.

71 PFCA, Project Dusting – Wigwam Inn 1929, 6 August 1929, G.R. Hopping to A.E. Parlow.

72 BCFSA, File 048811 – MFP Insects, 19 August 1929, G.R. Hopping to A.E. Parlow, enclosing 7 August 1929, "Preliminary Report of Airplane Dusting at Wigwam Inn, Burrard Inlet, British Columbia," George R. Hopping, from which the citations are taken.

73 Ibid.

74 Hopping, "The Western Hemlock Looper," 36–7.

75 PFCA, Project Dusting – Wigwam Inn 1929, 5 August 1929, R. Hopping to G.R. Hopping.

76 BCFSA, File 048811 – MFP Insects, 25 July 1929, W.E. Crosby to British Columbia Forest Service; 30 July 1929, E.B. Prowd to W.E. Crosby; and 5 August 1929, R. Hopping to Proud [sic].

77 PFCA, Project Dusting – Wigwam Inn 1929, 30 August 1929, J.H. Boyd to P.Z. Caverhill; 5 September 1929, E.B. Prowd to J.H. Boyd; and 5 September 1929, E.B. Prowd to G.R. Hopping.

78 PFCA, Project Dusting – Wigwam Inn 1929, 7 August 1929, E. Walmsley to G. Hopping and 10 August 1929, E. Walmsley to G.R. Hopping; LAC, RG39, 341, 47278, 10 August 1929, E. Walmsley to G.R. Hopping.

79 LAC, RG39, 341, 47278, 7 August 1929, L. Walmsley to C. MacFayden.

80 Even the project's loudest proponents had to admit that the operation had done nothing to control the local infestation: BCFSA, File 048811 – MFP Insects, 12 August 1929, A.E. Parlow to P.Z. Caverhill; LAC, RG39, 341, 47278, 17 August 1929, E. Walmsley to E.H. Finlayson.

81 LAC, RG39, 341, 47278, 12 August 1929, C. MacFayden to E.H. Finlayson.

82 Ibid.

83 LAC, RG39, 341, 47278, 24 August 1929, J.M. Swaine to D.R. Cameron.

84 LAC, RG39, 341, 47278, 27 August 1929, E. Walmsley to D.R. Cameron.

85 PFCA, Project Western Hemlock Looper 1929–1930, 31 August 1929, G.R. Hopping to R. Hopping; PFCA, Estimates [1929–1930] – 2012-08-01-019, "Estimates" [for 1929–1930], from which the citations are taken; LAC, RG39, 341, 47278, 27 August 1929, E. Walmsley to D.R. Cameron.

6 "Carrying out this work, of a protective nature"

1 PFCA, Estimates [1929–1930] – 2012-08-01-019, "Estimates." See also PFCA, Project Western Hemlock Looper 1929–1930, 31 August 1929, G.R. Hopping to R. Hopping.

2 White, *The Organic Machine*, passim; Schneider, *Hybrid Nature*, passim.

3 PFCA, FIDS Annual Reports 1923–1962, 10 October 1929, Report on Hemlock Looper and Other Insect Investigations in British Columbia, G.R. Hopping.

4 Ibid.

5 S. Kheraj, *Inventing Stanley Park*, passim.

6 CVA, VPK, S97, MSS50-F-1, Park Board – Clippings, *Christian Science Monitor*, April 16, 1926, "Stanley Park Scenic Beauty Attracts Tourists of the World."

7 CVA, VPK, S97, MSS50-F-1, Park Board – Clippings, *Vancouver Daily Province*, October 20, 1925, "Vancouver Wins Appeal Against Park Squatters," and articles from April–May 1926 regarding "the war on earwigs." See also CVA, VPK, S97, MSS50-F-1, Park Board – Clippings, *Vancouver Daily Province*, November 11, 1928, "W.S. Rawlings Responds to 'Nature Lover'"; and March 15, 1929, "Fishing at Beaver Lake and Lagoon."

8 D.R. MacLaren, "Dusting Defoliated Areas," *The Bulletin* [Canadian Airways Limited] 3, no. 10 (15 April 1932).

9 The early history of Vancouver's water supply and the aerial dusting of the Seymour Canyon in 1930 is retold in detail in Kuhlberg, "'An Eden that is practically uninhabited by humans.'" This paragraph, and the following ones that lack specific references, are based upon this source.

10 "Greater Vancouver Board Plans Water Supply for 3,000,000 People," *Vancouver Daily Province*, June 26, 1927.

11 "Water Supply Is Big Asset," *Vancouver Sun*, May 13, 1926.

12 "Not Exactly Capilano," *Vancouver Daily Province*, September 18, 1926.

13 British Columbia Legislative Library, OC1601/1918 and OC813/1923, with the citations taken from the first document.

14 BCA, GR-1441, 04214 #2, 10 October 1929, E.B. Prowd to R. Hopping.

15 BCA, GR-1441, 04214 #2, 23 October 1929, R. Hopping to E.B. Prowd.

16 BCFSA, File 048811 – MFP Insects, 1 November 1929, J.R. Grant to P.Z. Caverhill; 15 November 1929, P.Z. Caverhill to J.R. Grant; 9 November 1930, E.B. Prowd to A.E. Parlow; and 13 November 1929, A.E. Parlow to E.B. Prowd. See also LAC, RG39, 341, 47278, 21 November 1929, C. MacFayden to E.H. Finlayson.

17 PFCA, Project Dusting – Wigwam Inn 1929, 6 November 1929, J.M. Swaine to G. Hopping. See also PFCA, Project Dusting – Wigwam Inn 1929, 11 October 1929, J.J. de Gryse to G.R. Hopping.

18 PFCA, Project Dusting – Wigwam Inn 1929, 14 November 1929, G.R. Hopping to J.M. Swaine.

19 PFCA, Project Dusting – Wigwam Inn 1929, 28 November 1929, J.M. Swaine to G.R. Hopping.

20 PFCA, Project Dusting – Wigwam Inn 1929, 19 November 1929, G.R. Hopping to J.M. Swaine.

21 LAC, RG39, 341, 47278, 21 November 1929, C. MacFayden to E.H. Finlayson. Officials from the
    BCFB reminded the gathering that the province had already implemented legislation in the early
    1920s that obliged timber licensees to pay for half the cost of all control work done to contain
    beetle infestations on the areas they leased. There was some discussion of both amending the
    statute to make it applicable to aerial dusting operations and introducing a similar law at the
    federal level, but nothing came of these talks: BCFSA, File 048811 – MFP Insects, 23 November
    1929, R.C. St. Clair to P.Z. Caverhill.
22 LAC, RG39, 341, 47278, 21 November 1929, C. MacFayden to E.H. Finlayson.
23 LAC, RG39, 341, 47278, 31 December 1929, E. Walmsley to E.H. Finlayson.
24 LAC, RG39, 341, 47278, 15 November 1929, C. MacFayden to E.H. Finlayson; 3 December 1929,
    D.R. Cameron to District Forest Inspector, BC; and 31 December 1929, E. Walmsley to G.R.
    Hopping. See also PFCA, Project Dusting – Wigwam Inn 1929, 30 November 1929, J.M. Swaine
    to G.R. Hopping.
25 BCFSA, File 048811 – MFP Insects, 23 November 1929, R.C. St. Clair to P.Z. Caverhill, with
    St. Clair paraphrasing Powell's views.
26 Hopping, "The Western Hemlock Looper."
27 PFCA, Project Western Hemlock Looper 1929–1930, 28 November 1929, G.R. Hopping to
    J.M. Swaine. See also PFCA, Project Western Hemlock Looper 1929–1930, 31 May 1930, G.R.
    Hopping to J.M. Swaine.
28 PFCA, Project Western Hemlock Looper 1929–1930, 17 October 1929, G.R. Hopping to J.M.
    Swaine.
29 PFCA, Project Dusting – Wigwam Inn 1929, 19 November 1929, G.R. Hopping to J.M. Swaine.
30 PFCA, Project Dusting – Wigwam Inn 1929, 28 November 1929, J.M. Swaine to G.R. Hopping.
31 Ibid.
32 PFCA, Project Western Hemlock Looper 1929–1930, 18 December 1929, J.M. Swaine to G.R.
    Hopping. As a result, Hopping had already begun ruminating over new hopper designs, and
    Swaine offered as much advice as he could on the subject.
33 LAC, RG39, 341, 47278, ca. November 1929, Preliminary Report on Airplane Dusting in
    Manicouagan, J.M. Swaine.
34 PFCA, FIDS Annual Reports 1923–1962, 26 April 1930, Report on British Columbia Coast
    Insect Investigations, G.R. Hopping.
35 CVA, VPK, S97, MSS50-F-1, Vancouver Park Board Clippings, December 13, 1929, *Vancouver
    Star*, "Aerial Dusting Urged to Save Stanley Park."
36 See also CVA, VPK, S97, MSS50-F-1, Vancouver Park Board Clippings, December 13, 1929,
    Vancouver Sun, "Stanley Park Tree Pest."
37 CVA, VPK, S1, MSS49-C-7, 17 – Parks Board Annual Reports, 31 December 1929,
    *Superintendent's Annual Report and Resume of Work of Park System for the Year Ending December
    31st, 1929*, W.S. Rawlings.
38 PFCA, Project Dusting Stanley Park Seymour Watershed 1930, 5 April 1930, G.R. Hopping to
    W.S. Rawlings.
39 PFCA, Annual FIDS Coastal Region BC 1929, 1 April 1930, Annual Report, R. Hopping,
    Entomology.
40 PFCA, Project Dusting Stanley Park Seymour Watershed 1930, 24 April 1930, W.S. Rawlings to
    G.R. Hopping. See also CVA, VPK, S76, MSS48-A-3, Park Board Minutes 1928–1930, Minutes
    from Meeting on 17 April 1930.
41 CVA, VPK, S97, MSS50-F-1, Vancouver Park Board Clippings, April 23, 1930, *Vancouver Star*,
    "Loss of Trees Inevitable, If Nothing Done."

42 CVA, VPK, S97, MSS50-F-1, Vancouver Park Board Clippings, 24 April 1930, *Vancouver Daily Province*, "The Looper." See also CVA, VPK, S97, MSS50-F-1, Vancouver Park Board Clippings, April 18, 1930, *Vancouver Star*, "Board Lacks Cash to Prevent Tree Destruction"; April 23, 1930, *Vancouver Star*, "Save the Trees? Of Course!"; and April 24, 1930, *Vancouver Star*, "Up to Council to Find Cash to Kill Looper, Says Crowne."

43 PFCA, FIDS Annual Reports 1923–1962, 26 April 1930, Report on British Columbia Coast Insect Investigations, G.R. Hopping. See also, PFCA, Project Dusting Stanley Park Seymour Watershed 1930, 26 April 1930, G.R. Hopping to W.S. Rawlings; 26 April 1930, G.R. Hopping to J.R. Grant; and 1 May 1930, W.S. Rawlings to G.R. Hopping; CVA, VPK, S76, MSS48-A-3, Parks Board Minutes 1928–1930, Minutes from Meeting on 1 May 1930.

44 PFCA, Project Dusting Stanley Park Seymour Watershed 1930, 1 May 1930, W.S. Rawlings to G.R. Hopping.

45 PFCA, Project Dusting Stanley Park Seymour Watershed 1930, 2 May 1930, G.R. Hopping to W.S. Rawlings. Word nevertheless leaked out about the Board's decision soon thereafter, but the criticism it feared from the media did not materialize: see CVA, VPK, S97, MSS50-F-1, Vancouver Park Board Clippings, May 26, 1930, *Vancouver Star*, "Park Trees to Be Saved From Looper"; and June 2, 1930, *Vancouver Star*, "Extermination of Looper to Be Undertaken."

46 PFCA, Annual FIDS Coastal Region BC 1929, 1 April 1930, Annual Report, R. Hopping, Entomology.

47 PFCA, Project Dusting Stanley Park Seymour Watershed 1930, 24 April 1930, J.R. Grant to G.R. Hopping.

48 PFCA, Project Dusting Stanley Park Seymour Watershed 1930, 14 April 1930, G.R. Hopping to J.R. Grant; 14 April 1930, J.R. Grant to G.R. Hopping; 26 April 1930, G.R. Hopping to J.R. Grant; 2 May 1930, J.R. Grant to G.R. Hopping; 19 May 1930, G.R. Hopping to J.R. Grant; and 20 May 1930, per J.R. Grant to G.R. Hopping

49 PFCA, FIDS Annual Reports 1929–1962, 10 May 1930, Report on British Columbia Insects, G.R. Hopping; and 24 May 1930, Report on British Columbia Insects, G.R. Hopping.

50 PFCA, Project Dusting Stanley Park Seymour Watershed 1930, 3 May 1930, E.A. Cleveland to G.R. Hopping.

51 Hopping believed that it would not be possible to treat the infestation in the upper part of the canyon because its walls were too steep: CVA, MSS[AM]1257, S13, 64-B-06, 6 – Hemlock Looper 1929–1931, 7 April 1931, E.A. Cleveland to C.S. Cowan.

52 PFCA, Project Dusting Stanley Park Seymour Watershed 1930, 2 May 1930, G.R. Hopping to Greater Vancouver Water Board; and 9 May 1930, G.R. Hopping to C.E. Cleveland. See also PFCA, FIDS Annual Reports 1929–1962, 24 May 1930, Report on British Columbia Insects, G.R. Hopping; and PFCA, Project Western Hemlock Looper 1929–1930, 19 May 1930, G.R. Hopping to "Dear Ray."

53 PFCA, Project Dusting Stanley Park Seymour Watershed 1930, 12 May 1930, G.R. Hopping to Greater Vancouver Water Board.

54 PFCA, Project Dusting Stanley Park Seymour Watershed 1930, 13 May 1930, E.A. Cleveland to G.R. Hopping; and PFCA, FIDS Annual Reports 1929–1962, 24 May 1930, Report on British Columbia Insects, G.R. Hopping.

55 PFCA, Project Dusting Stanley Park Seymour Watershed 1930, 17 March 1930, J.M. Swaine to G.R. Hopping; and 12 April 1930, J.M. Swaine to G.R. Hopping. It is unclear why Swaine did not permit the trimotor Ford to be used in BC that year, but undoubtedly he was averse to the aircraft logging the thousands of miles it would have had to have flown from Ottawa to Vancouver, and back.

56 LAC, RG24-E-1-a, 4913, DND – Operations for Department of Agriculture, vol. 4, 3 April 1930, J.M. Swaine to J.L. Gordon; and 17 April 1930, J.H. Grisdale to J.L. Gordon; PFCA, Project Dusting Stanley Park Seymour Watershed 1930, 3 and 16 May 1930, J.M. Swaine to G.R. Hopping; 10 May 1930, J.M. Swaine to G.R. Hopping, enclosing 8 May 1930, Chief Aeronautical Engineer, Memorandum, "Dusting Hopper"; and 17 May 1930, J.M. Swaine to G.R. Hopping [telegram].

57 PFCA, Project Dusting Stanley Park Seymour Watershed 1930, 21 May 1930, G.R. Hopping to J.M. Swaine.

58 PFCA, Project Dusting Stanley Park Seymour Watershed 1930, 30 May 1930, J.M. Swaine to G.R. Hopping. See also PFCA, Project Dusting Stanley Park Seymour Watershed 1930, 4 June 1930, G.R. Hopping to J.M. Swaine; LAC, RG24-E-1-a, 4913, DND – Operations for Department of Agriculture, Vol. 5, 2 May 1930, J.L. Gordon to J.M. Swaine; and 8 May 1930, A. Ferrier to C.O.F.

59 G.R. Hopping, "Dusting by Airplane in British Columbia," *The Timberman* 32, no. 3 (January 1931): 24.

60 Reeves, "In Stanley Park, You Say?"

61 PFCA, Project Dusting Stanley Park Seymour Watershed 1930, 16 June 1930, J.M. Swaine to G.R. Hopping. See also PFCA, Project Dusting Stanley Park Seymour Watershed 1930, 7 April 1930, G.R. Hopping to J.M. Swaine; 17 April 1930, J.M. Swaine to G.R. Hopping; 16 May 1930, J.M. Swaine to G.R. Hopping; and 23 May 1930, J.M. Swaine to G.R. Hopping.

62 PFCA, FIDS Annual Reports 1923–1962, Insect Problems of the British Columbia Coast, Season of 1930, G.R. Hopping; 10 May 1930, Report on British Columbia Insects, G.R. Hopping; 24 May 1930, Report on British Columbia Insects, G.R. Hopping; and 10 June 1930, Report on British Columbia Insects, G.R. Hopping.

63 PFCA, Project Dusting Stanley Park Seymour Watershed 1930, 8 May 1930, Oliver Chemical Company to G.R. Hopping; and 24 and 28 May 1930, C.E. Oliver to G.R. Hopping.

64 PFCA, Project Dusting Stanley Park Seymour Watershed 1930, 8 May 1930, G.R. Hopping to J.M. Swaine; 21 May 1930, G.R. Hopping to W.S. Rawlings; 6 June 1930, L.A. Dobbin to G.R. Hopping; and 12 June 1930, D.R. MacLaren to G.R. Hopping. CVA, VPK, S97, MSS50-F-1, Vancouver Park Board Clippings, June 12, 1930, *Vancouver Star*.

65 PFCA, FIDS Annual Reports 1923–1962, Insect Investigations on B.C. Coast – 1930 – Airplane Dusting on the British Columbia Coast, 1930.

66 Ibid. See also CVA, MSS[AM]1257, S13, 64-B-06, Hemlock Looper 1929–1931, 7 June 1930, E.A. Cleveland to D.R. MacLaren; and 9 June 1930, enclosing "Airplane Dusting – Seymour Canyon 1930," G.R. Hopping.

67 PFCA, FIDS Annual Report 1923–1962, Insect Investigations on B.C. Coast – 1930 – Airplane Dusting on the British Columbia Coast, 1930.

68 PFCA, Project Dusting Stanley Park Seymour Watershed 1930, 26 June 1930, E.A. Cleveland to G.R. Hopping; CVA, VPK, S1, PDS-12-1-609, Annual Report of the Vancouver Board of Park Commissioners 1929–1931, *Superintendent's Resume of Work of Park System for the Year Ending December 31st 1930*, W.S. Rawlings; and *Superintendent's Resume of Work of Park System for the Year Ending December 31st 1931*, W.S. Rawlings.

69 CVA, MSS[AM]1257, S13, 64-B-06, Hemlock Looper 1929–1931, 29 July 1930, G.R. Hopping to E.A. Cleveland; see also 31 July 1939, E.A. Cleveland to G.R. Hopping.

70 CVA, MSS[AM]1257, S13, 64-B-06, Hemlock Looper 1929–1931, 26 June 1930, E.A. Cleveland to G.R. Hopping.

71 CVA, VPK, S76, MSS48-A-3, Parks Board Minutes 1928–1930, Minutes from Meeting on 10 July 1930; PFCA, Project Dusting Stanley Park Seymour Watershed 1930, 13 August 1930, G.R.

Hopping to District Forester for Vancouver, from which the citation is taken; 13 August 1930, G.R. Hopping to D.R. MacLaren; 13 August 1930, R.C. St. Clair to G.R. Hopping. Swaine had asked Hopping to explain how he had determined the effectiveness of the chemical dust, for he was most curious to learn an accurate means for doing so. Significantly, Hopping never replied directly to that specific question: PFCA, Project Dusting Stanley Park Seymour Watershed 1930, 5 August 1930, J.M. Swaine to G.R. Hopping.

72 CVA, MSS[AM]1257, S13, 64-B-06, Hemlock Looper 1929–1931, 29 July 1930, G.R. Hopping to E.A. Cleveland; see also 31 July 1930, E.A. Cleveland to G.R. Hopping.

73 PFCA, Project Dusting Stanley Park Seymour Watershed 1930, 9 July 1930, G.R. Hopping to W.S. Rawlings. See also PFCA, FIDS Annual Reports 1923–1962, Insect Investigations on B.C. Coast – 1930 – Airplane Dusting on the British Columbia Coast, 1930.

74 PFCA, Project Dusting Stanley Park Seymour Watershed 1930, 23 July 1931, J.M. Swaine to G.R. Hopping [extract].

75 PFCA, FIDS Annual Reports 1923–1962, Insect Problems of the British Columbia Coast, Season of 1930, G.R. Hopping.

76 PFCA, Annual FIDS Coastal Region BC 1929, 1 April 1930, Annual Report, R. Hopping, Entomology.

77 PFCA, Project Dusting Stanley Park Seymour Watershed 1930, 8 July 1930, J.M. Swaine to G.R. Hopping.

78 Castonguay, Protection des cultures, 116. See also PFCA, Project Western Hemlock Looper 1929–1930, 5 and 12 December 1929, J.M. Swaine to G.R. Hopping; 7 and 17 December 1929 and 2 July 1930, G.R. Hopping to J.M. Swaine; 8 March 1930, H.T. Gussow to J.M. Swaine; 13 March 1930, J.M. Swaine to G.R. Hopping; and 16 July 1930, G.S. Walley to G.R. Hopping.

79 PFCA, FIDS Annual Reports 1923–1962, Insect Investigations on B.C. Coast – 1930 – Airplane Dusting on the British Columbia Coast, 1930. See also PFCA, Project Dusting Stanley Park Seymour Watershed 1930, 13 August 1930, G.R. Hopping to J.M. Swaine.

80 PFCA, Project Dusting Stanley Park Seymour Watershed 1930, 15 November 1930, F.P. Keen to G.R. Hopping.

81 G.R. Hopping, "Dusting by Airplane in British Columbia," The Timberman 32, no. 3 (January 1931).

82 University of Washington Libraries – Special Collections, MSS1311-2 –Washington Forest Protection Association Records, Washington Forest Fire Association, Twenty-Third Annual Report 1930, 10; World Forestry Center, "Charles S. Cowan," 28–30.

83 BCA, GR-1441, 04214 #2, 16 March 1931, C.S. Cowan to P.Z. Caverhill; and 19 March 1931, E.B. Prowd to C.S. Cowan; CVA, MSS[AM]1257, S13, 64-B-06, Hemlock Looper 1929–1931, 24 March 1931, C.S. Cowan to E.A. Cleveland; and 26 March 1931, E.A. Cleveland to C.S. Cowan; PFCA, Airplane Dusting – Correspondence 1931–1932, 30 March 1931, G.R. Hopping to C.S. Cowan.

84 BCFSA, File 048811 – MFP Insects, 18 September 1930, P.M. Barr to R. Hopping; and 23 September 1930, R. Hopping to P.M. Barr.

85 BCFSA, File 048811 – MFP Insects, documents from 1932 to 1953, particularly those dealing with black-headed budworm.

86 BCFSA, File 048811 – MFP Insects, 16 July 1931, C.R. Richardson to P.Z. Caverhill. See also BCFSA, File 048811 – MFP Insects, 6 July 1931, Report on Insect Damage – Campbell River Timber Company Holdings – Garrett, British Columbia; BCFSA, File 048811 – MFP Insects, 21 July 1931, E.B. Prowd to C.R. Richardson; and 21 July 1931, E.B. Prowd to G.R. Hoppping.

87 BCFSA, File 048811 – MFP Insects, 23 July 1931, G.R. Hopping to E.B. Prowd.

88 BCFSA, File 048811 – MFP Insects, 24 July 1931, C.R. Richardson to G.R. Hopping.

89 BCFSA, 048811 Vol. 2, 27 October 1937, T.R. Swanson to C.D. Orchard.

90 BCFSA, 048811 Vol. 2, 1 November 1937, C.D. Orchard to T.R. Swanson. See also BCA, 04214 #4, documents from 1945.

Conclusion

1   LAC, RG24-E-1-a, 4913, DND – Operations for Department of Agriculture, Vol. 4, "Preliminary Report Airplane Dusting – Manicouagan 1929," J.M. Swaine.

2   LAC, RG24-E-1-a, 4913, DND – Operations for Department of Agriculture, Vol. 4, 17 April 1930, J.H. Grisdale to J.L. Gordon, enclosing "Airplane Dusting Operations for the Control of Defoliating Insects – Conducted in Cooperation with the Dominion Air Service in 1929," J.M. Swaine; see also Swaine, *The Spruce Budworm* and *The Hemlock Looper*; and SMPA, A-1, Forestry 1932, March 1933, Report of Organization for Forest Protection and Fire Data for the Sudbury Division of the Abitibi Power & Paper Company Limited, 1932, 34.

3   LAC, Vol. 6, 21 August 1932, State Research Institute of Agricultural Aviation, USSR to Department of National Defence, Canada.

4   LAC, RG39, 341, 47278, 31 January 1934, D.R. Cameron to J.M. Swaine, enclosing 19 January 1934, G.P. Vanier to Director, Dominion Forest Service.

5   Even though the Canadian forest entomologists were leagues ahead of their American colleagues in terms of aerial dusting, tellingly Vancouverites who were concerned about pests attacking Stanley Park's trees during this period wrote to prominent American forest entomologists for advice about how best to deal with the problem. The Americans quickly redirected the inquiries to the Canadians, whom the Americans felt were the best sources for this type of information. PFCA, Project Dusting Stanley Park Seymour Watershed 1930, 12 May 1930, J.M. Swaine to G.R. Hopping, enclosing 2 May 1930, C.E. Tisdall to US Entomologist, Washington, DC; and PFCA, Project Dusting Stanley Park Seymour Watershed 1930, 19 May 1930, G.R. Hopping to C.E. Tisdall.

6   CAS, RHP, 325, C.J. Drake, 28 January 1930, G.R. Hopping to C.J. Drake.

7   PFCA, Project Dusting Stanley Park Seymour Watershed 1930, 5 December 1930, F.P. Keen to G.R. Hopping; see also FHSA, F16.1 – USFS History Collection, 16 June 1930, "Forest Entomology in the Western US," F.P. Keen [A Paper Presented at A Symposium on Forest Trees at June 1930 Meeting of the Pacific Division of the American Association for the Advancement of Science]; BCPA, GR-1441, 04214 #2, 16 March 1931, C.S. Cowan to P.Z. Caverhill; and BCPA, GR-1441, 04214 #2, 19 March 1931, E.B. Prowd to C.S. Cowan. Hopping published most of his findings in G.R. Hopping, "An Account of the Western Hemlock Looper, *Ellopia somniaria* Hulst, on Conifers in British Columbia," *Scientific Agriculture* 15, no.1 (September 1934): 12–29.

8   LAC, RG24-E-1-a, 4913, DND – Operations for Department of Agriculture, Vol. 4, 11 September 1929, J.M. Swaine to J.L. Gordon; 8 November 1929, A. Gibson to J.H. Grisdale; and 12 December 1929, G.J. Desbarats to J.H. Grisdale.

9   LAC, RG24-E-1-a, 4913, DND – Operations for Department of Agriculture, Vol. 4, 22 November 1930, J.H. Grisdale to G.J. Desbarats; 2 March 1931, A. Gibson to J.H. Grisdale; and fall 1931 correspondence regarding aerial survey of Gaspe peninsula.

10 LAC, RG17, 3131, 69 – Department of Militia and Defence, 26 October 1929, G.J. Desbarats to J.H. Grisdale; 30 October 1929, J.H. Grisdale to A. Gibson and H.T. Gussow; and 8 November 1929, A. Gibson to J.H. Grisdale; LAC, RG24-E-1-a, 4913, DND – Operations for Department of Agriculture, Vol. 5, all documents; Richmond, "A History of Forest Entomology in British Columbia," 5; Rajala, "The Vernon Laboratory," 183.

11 Beginning as early as 1928, Swaine had stressed the need for all parties who were involved in Canadian forestry to cooperate in generating information about the pests that were damaging the country's woodlands. Within a few years, he had set up the unofficial network that would grow to form the backbone of the forest insect intelligence service. This was a system whereby Swaine's Division of Forest Insects would educate foresters across the country in the type of information that the foresters could collect that would be most helpful in tracking and learning about insects; his office even provided blank forms on which the data could be recorded. The network would soon be overseen by J.J. de Gryse, Swaine's colleague, and it grew to include officials from the national and provincial forest services, the Canadian Parks Branch, lumber and pulp and paper firms, industry lobby groups, and forest entomologists. The information it obtained each year would be published annually as the Canadian Forest Insect Survey, which serves as a remarkable entomological record. J.M. Swaine, "Progress in Forest Insect Control in Canada," *Forestry Chronicle* 4, no. 1 (1928): 40; Gibson, "The Canadian Entomological Service," 1454–5.

12 AO, RG1–256, 1, Hemlock Looper (1), [?] August 1927, M. Cornell to Canadian Department of Agriculture. See also, AO, RG1–256, 1, Hemlock Looper (1), 18 October 1927, H.W. Crosbie to J.J. de Gryse [2]; UTA, B83–0022/2, File 256, 7 October 1927, H.W. Crosbie to J.M. Swaine; Canada Department of Agriculture, Entomological Branch, *The Canadian Insect Pest Review* 5, no. 4 (August 1927): 42; J.J. de Gryse [no title], (Ottawa: Dominion Entomological Branch, 1928), unpublished but cited in de Gryse and K. Schedl, "An Account of the Eastern Hemlock Looper, *Ellopia fiscellaria* Gn., on Hemlock, with Notes on Allied Species," *Scientific Agriculture* 14, no. 10 (1934).

13 UTA, B83–0022, File 256, 13 December 1927, H.W. Crosbie to J.H. White, enclosing 7 October 1927, H.W. Crosbie to J.M. Swaine; and 11 October 1927, J.J. de Gryse to H.W. Crosbie. See also AO, RG1–256, 1, Hemlock Looper (1), 1 December 1927, J. Fitzpatrick to E.H. Finlayson; 2 December 1927, E.H. Finlayson to J. Fitzpatrick; 2 December 1927, E.H. Finlayson to E.J. Zavitz; and 16 December 1927, E.J. Zavitz to J. Fitzpatrick

14 Canada Department of Agriculture, Entomological Branch, *The Canadian Insect Pest Review* 6, no. 5 (October 1928): 46. See also Department of Agriculture, Entomological Branch, *The Canadian Insect Pest Review* 6, no. 2 (June 1928): 17.

15 Whorton, "Insecticide Spray Residues and Public Health"; Whorton, *Before Silent Spring*.

16 Bastedo, *Materia Medica*.

17 Swaine and Craighead, *Studies on the Spruce Budworm*, 88. There is a striking contrast between the Canadian and American perspectives on this issue. Around the same time as Swaine's budworm report was published, the United States Department of Agriculture released a report in which it boasted of the smashing success it was enjoying in aerial dusting against pests that were harming cotton crops. Significantly, it downplayed the potential collateral damage of such efforts but provided no concrete evidence upon which to base its assessment: see Coad et al., *Dusting Cotton from Airplanes*.

18 LAC, RG39, 341, 47278 – Head Office – Pests – Dusting Experiments, 18 November 1926, The Inspector of Forests at Haguenau to the Director of the International Company of Aerial Navigation, Care of the Chief of the Entzheim Center [translation]. These officials soon stopped using the masks, and the French concluded that the calcium arsenate was "practically inoffensive of game and birds."

19 LAC, RG24-E-1-a, 4912, DND – Operations for Department of Agriculture, Vol. 2, 25 April 1927, B.R. Coad to J.A. Wilson.

20 LAC, RG24-E-1-a, 4912, DND – Operations for Department of Agriculture, Vol. 2, 29 April 1927, T.M. Shields to CO, Borden. See also LAC, RG24-E-1-a, 4912, DND – Operations for

Department of Agriculture, Vol. 2, 29 April 1927, C.L. Bath, Report on "Dusting Courses." Although Squadron Leader A.E. Godfrey placed an order for "Special Goggles for Dusting," his superiors did not purchase the correct or even effective ones: LAC, RG24-E-1-a, 4912, DND – Operations for Department of Agriculture, Vol. 2, 19 May 1927, A.E. Godfrey, Memorandum re: Special Goggles for Dusting.

21 PFCA, Project Dusting Stanley Park Seymour Watershed 1930, 30 May 1930, J.M. Swaine to G.R. Hopping.

22 PFCA, Project Dusting Stanley Park Seymour Watershed 1930, 4 June 1930, G.R. Hopping to J.M. Swaine.

23 LAC, RG24-E-1-a, 4913, DND – Operations for Department of Agriculture Vol. 5, 23 April 1931, A.E. Godfrey to Director CGAO.

24 LAC, RG24-E-1-a, 4913, DND – Operations for Department of Agriculture Vol. 5, 28 April 1931, D.E. Dewar to DCO.

25 LAC, RG39, 341, 47278, 18 November 1926, The Inspector of Forests at Haguenau to the Director of the International Company of Aerial Navigation, c/o of the Chief of the Entzheim Center [translation]; LAC, RG39, 341, 47278, 16 March 1928, J.M. Swaine to D.R. Cameron; LAC, RG24-E-1-a, 4912, DND – Operations for Department of Agriculture Vol. 3, 10 February 1928, Secretary to British Air Attache in Berlin to E.W. Stedman, enclosing "Fighting of Pests."

26 AO, RG1–256, 1, Hemlock Looper (2), 19 October 1928, "Summary Report on Airplane Dusting Operations, Foote's Bay, Ontario, 1928," J.J. de Gryse.

27 PFCA, Project Dusting Stanley Park Seymour Watershed, 21 March 1930, J.M. Swaine to G. Hopping.

28 Ibid.

29 PFCA, Project Dusting Stanley Park Seymour Watershed, 21 March 1930, J.M. Swaine to G. Hopping. See also Prebble, *Aerial Control of Forest Insects in Canada*, 8. What is all the more surprising about the lack of concern the Canadians expressed about the collateral damage the dusts caused was the evidence that strongly suggests that this issue was relatively important at the time. When the Americans went to conduct their aerial dusting project in Washington State in 1931, F.P. Keen, one of the most senior forest entomologists within the US Department of Agriculture's Bureau of Entomology, had voiced his concern over this very subject to F.C. Craighead, the Bureau's chief. Keen raised the matter because he explained that the planned project was taking flak over the potential impact it might have on wildlife and domestic animals. He explained that he had attempted to obtain information from the local health boards about what constituted a lethal dose of calcium arsenate for birds, mammals, and humans, but his effort had been in vain: FHSA, F16.1 – USFS History Collection, Wickman, Torgensen, and Furniss, "Photographic Images and History of Forest Investigations."

30 PFCA, Project Dusting – Wigwam Inn 1929, 6 July 1929, A.A. Dennys to G.R. Hopping.

31 PFCA, Project Dusting – Wigwam Inn 1929, 28 November 1929, J.M. Swaine to G.R. Hopping.

32 PFCA, Project Dusting – Wigwam Inn 1929, 3 December 1929, J.M. Swaine to G.R. Hopping, in which Swaine quotes Shutt.

33 Carson, *Silent Spring*; Dunlap, *DDT*; Whorton, *Before Silent Spring*; Russell, *War and Nature*; Mart, *A Love Story*.

34 McLaughlin, "Green Shoots," 4. See also Armstrong and Cook, *Aerial Spray Applications on Canadian Forests*; Sandberg and Clancy, "Politics, Science and the Spruce Budworm."

35 LAC, RG34, 255, Mosquito Control 1940s – DDT, all documents; LAC, RG17-B-V-1, 4338, 3802 – FDA – DDT, 1944–1956, all documents; Turner, "A History of Investigations"; Armstrong and Cook, *Aerial Spray Applications on Canadian Forests*; Prebble, *Aerial Control of Forest Insects in Canada*.

36 Loo, *States of Nature*, 2.

Bibliography

Primary Sources

ARCHIVES

Abitibi-Bowater Archives – Iroquois Falls Division (now closed)
  Forest History Files.

Ahmic Lake Cottagers Association
  Ahmic Lake Cottage Owners' Association – Minutes of Annual Meetings, 1928–1929.

Archives of Ontario (AO)
  <u>Government Collections</u>
  RG1 – Department of Lands and Forests/Ministry of Natural Resources, Series 116, 256, 448, and 560.
  RG17 – Archives of Ontario – Historical Research Files, Series 20.
  <u>Private Collections</u>
  F229–37 – Destructive Insect and Pest Act, 1928–1929.

British Columbia Archives (BCA)
  <u>Government Collections</u>
  GR-0111 – Provincial Museum correspondence inward, 1897–1970.
  GR-0132 – Public Health Officer, 1898–1957.
  GR-0385 – BC Department of Lands and Forests – Land records related to leases, 1865–1955.
  GR-0436 – Land settlement records for Railway Belt and Peace River Block, 1885–1949.
  GR-0441 – Premier's Papers, Simon Fraser Tolmie.
  GR-0520 – Commission on Forest Resources, 1943–1945.
  GR-0943 – BC Forest Branch – Timber survey files, 1912–1943.
  GR-0946 – BC Forest Branch – Timber investigations files, 1913–1915.

GR-0948 – BC Forest Branch – Executive Records, 1915–1920.

GR-0955 – Vancouver Forest District – Operation Records, 1912–1979.

GR-0993 – Correspondence index to lands files.

GR-1190 – BC Plant Pathologist, Entomology Branch, 1915–1933.

GR-1209 – BC Forest Branch, Chief Forester's Correspondence, 1919–1929.

GR-1230 – BC Forest Branch – Forest Research Branch, 1929.

GR-1365 – Index to Forestry Correspondence, 1918–1980s.

GR-1384 – Railway Belt land leases, 1911–1946.

GR-1441 – Correspondence files ("O" Series) with regard to Crown lands, Files 04214, 04215, 04216, 04217, 04221, 04220, 04937, 0120161.

GR-3100 – Timber and pulp licence registers, 1861–1953.

Private Collections

MSS-0003 – D. Pattullo Papers.

MSS-0035 – M.A. Grainger Papers.

MSS-0588 – M.A. Grainger Papers.

MSS-0840 – C.D. Orchard Papers.

MSS-1302 – C.D. Orchard Papers.

British Columbia Forest Service Archives (BCFSA)

File 048811 – MFP Insects.

British Columbia Legislative Library

Orders-in-Council (OC).

California Academy of Sciences (CAS) [San Francisco, California]

Ralph Hopping Papers, 7 boxes (unprocessed).

City of Vancouver Archives (CVA)

Government Collections

MSS[AM]1257 – Greater Vancouver Regional District Fonds.

S9 – Greater Vancouver Water District Board minute books, 1926–1962.

S10 – Greater Vancouver Water District Board minutes, 1931–1952.

S13 – Greater Vancouver Water District Correspondence, 1926–1962: 64-B-06.

S15 – Greater Vancouver Water District Reports, 1922–1967.

S26 – Greater Vancouver Regional District reference maps.

COV – City of Vancouver

S31 – Vancouver City Council Minutes.

VPK – Vancouver Park Board

S1 – Board of Park Commissioners – Annual reports, 1911–1955: 48-A-2; 49-C-7; PDS-12–1.

S64 – Committee Minutes, 1912–1961: A-6.

S76 – Board Minutes, 1888–1987: MCR-47–1; MSS-48-A-3.

S81 – Correspondence, 1904–1967: 48-C-1; 48-C-5; 49-B-5; 49-C-7; 49-D-1.

S88 – Legal Agreements and Expense Reports, 1896–1958.

S92 – Disbursement Ledgers, 1913–1930: 54-C-2.

S97 – Newsclippings and Photographs, 1911–1957: 41-C-06; 50-E-4; 50-F-01.

S100 – Journals, 1912–1948.

Private Collections

AM54 – Major Matthews Collection.

Cornell University Archives (CUA) – Rare Books and Manuscripts
MSS1781 – West Virginia Pulp and Paper Company, Volumes 26 and 108.

Forest History Society Archives (FHSA) [Durham, North Carolina]
F16.1 – USFS History Collection, Biographical information on Ernst J. Schreiner.

Iowa State University – Special Collections and Archives [Ames, Iowa]
Hopping, G.R. "The Western Hemlock Looper." MSc thesis, Iowa State College, 1931.

Library and Archives Canada (LAC)
<u>Government Collections</u>
RG6 – Secretary of State, Series A-1.
RG13 – Department of Justice, Series A-2, F-6.
RG15 – Department of the Interior, Series, Series D-V-1.
RG17 – Department of Agriculture, Volumes 1107, 1098, 2814, 3037, 3040, 3041, 3042, 3043, 3044, 3047, 3050, 3051, 3131, 3159, 3163, 3331, Series A-II-7, B-VII-1a, B-VII-3-a.
RG24 – Department of National Defence, Series E-1-a, E-14.
RG25 – Department of External Affairs, Series EA, B-1-b.
RG39 – Dominion Forest Branch, Volumes 148, 248, 341, 416, 417, 418, 452.
RG88 – Surveys and Mapping Branch, Series B-4.
<u>Private Collections</u>
MG27-III-B11 – J.L. Ralston, Vol 13 – Personal Corr and political corr from 1923–30.
MG27-III-B25 – W.R. Motherwell.
MG28-I25 – Entomological Society of Canada.
MG28-I188 – Canadian Forestry Association.
MG30-B91 – E.G.D. Murray.
MG30-D337 – J.H. Mickelthwaite.
MG30-E89 – G.J. Desbarats.
MG30-E243 – J.A. Wilson.
MG30-E322 – E.W. Stedman.
MG30-E562 – K.M. Guthrie.
MG31-E97 – W.A. Irwin.
Mikan23982 – N.C. Ogilvie-Forbes (graphic material).
National Map Collection
R8706–0 - 8-E – D.A. Dwyer.
R12712–0 - 8-E – F.T. Jenkins.

Maine Historical Society [Portland, Maine]
Collection 1882.

National Film Board (NFB)
An aeroplane dusting experiment for the control of spruce bud worm carried out at Westree on [*sic*] June 1929.

Natural Resources Canada – Pacific Forestry Centre (PFCA) [Victoria, BC]
Airplane Dusting – Correspondence, 1931–1932.
Annual Forest Insect Investigations and Detailed Studies [FIDS] Coastal Region BC 1929.
Estimates [1929–1930].
Forest Insect Investigations and Detailed Studies [FIDS] Annual Reports 1923–1962.

Project Dusting Stanley Park Seymour Watershed 1930.
Project Dusting – Wigwam Inn 1929.
Project Western Hemlock Looper 1929–1930.
Unpublished Letters at NR Canada, PFC.

Nova Scotia Archives (NSA)
O/S V/F 28-10 – *Forests and Forestry in Nova Scotia, W.L. Hall and Otto Schierbeck – Special Report to the British Empire Forestry Conference, Australia 1928.*
RG10 – Department of the Attorney General, Series B.
RG20 – Department of Lands and Forests, Volumes 718–733, 742–749, 774–776, 826–837.

Smithsonian Institute – Archives, Washington, DC (SIA)
Record Unit [RU] 140 – United States National Museum, Division of Insects, 1908–1961.
Box 5, Craighead, F.C.
Box 11, McDunnough, J.H.
Box 12, McDunnough, J.H.
RU 7121, E.D. Ball
Box 2, Howard, L.O.
Box 3, McDunnough, J.H.
Box 4, Van Duzee, E.D.
RU 7103, H.S. Barber
Box 2, Blaisdell, F.E.
Box 4, Chittenden, F.H.; Craighead, F.C.
Box 6, Howard, L.O.
Box 10, Swaine, J.M.
RU 7310, D.H. Blake
Boxes 3–5, Hopping, R.

St. Mary's Paper Archives (SMPA) (now closed)
Filing Cabinet Drawer A-1.

United States National Archives and Records Administration (NARA) [Washington, DC]
RG95 – Office of the Chief Forester [United States Forest Service]
Box 5, F Reports – District Forester, 1914–1916.
Box 6, F Reports – District Forester, 1912–1914.
Box 9, F Supervision – Supervisors & Rangers Mtgs, 1915.
Box 29, F Supervision – Graves, Henry S. (two folders).
Box 34, F Supervision – Office Orders (Misc. (O) Series), 1915.

United States National Archives and Records Administration – Kansas City
RG7 – Records of the Bureau of Entomology and Plant Quarantine, Series – General Correspondence, 1925–1934.

United States National Parks Service, Harpers Ferry Centre [Charles Town, West Virginia]
Assembled Records of the National Park Service, Series XI: Themes, Activities and Events Records, 1892–2006, Sub-Series C: Conferences, 1911–2005, Box 12, Folder 5, *Proceedings of the National Park Conference, Held at Berkeley, California, March 11, 12 and 13, 1915.* Washington, DC: National Printing Office, 1915.

University of Idaho Library – Special Collections and Archives [Moscow, Idaho]
    MG134 – J.C. Evenden.
    Western Forest Insect Work Conference Archives.

University of Toronto Archives
    A1972–0025 – Records of the Faculty of Forestry.
    B83–0022 – J.H. White Papers.

University of Washington Libraries – Special Collections Division [Seattle, Washington]
    MSS1301 – Washington Forest Protection Association Records, 1908–1988.

University of Waterloo Archives (UWA)
    GA 100 – Muskoka Lakes Association, Series 1, 4, and 6.

West Virginia Historical Archives & Manuscript Collections – West Virginia University [Morgantown, West Virginia]
    A&M.0904 – Andrew Delmar Hopkins Papers.

Wisconsin Historical Society Archives [Madison, Wisconsin]
    PH6159 MAD 4/92/E2 – Photographs of Hemlock Looper Dusting, Door County, Wisc., 1926.
    Series 271 – Wisconsin Conservation Department: Subject Files, 1917–1968: Boxes 864 and 897.

Yale University Library – Manuscripts and Archives
    MS134, Herman Haupt Chapman, Series I, Box 44, File 135.
    MS249, Henry Solon Graves Papers.
        Series I, Box 8, File 88.
        Series II, Box 19, Files 223–227; Box 44, Files 133 and 135.

GOVERNMENT DOCUMENTS

Armstrong, J.A., and C.A. Cook. *Aerial Spray Applications on Canadian Forests: 1945 to 1990* (Information Report ST-X-2). Ottawa: Forestry Canada, 1993.

Canada. *Sessional Papers, 1908–1931*.

Canada Commission of Conservation. *First Annual Report 1910*. Ottawa: Mortimer Company, 1910.

Canada Department of Agriculture. *Annual Report of the Entomologist and Botanist for the Calendar Year 1886*. Ottawa: Maclean, Roger & Co., 1887.

– *Annual Reports of the Minister of Agriculture for the Dominion of Canada, 1925–1932*.

Canada Department of Agriculture, Entomological Branch. *The Canadian Insect Pest Review, 1923–1932*.

Canada Department of the Interior. *Natural Resources Canada* 6, no. 9 (September 1927).

Canada Department of National Defence. [*Annual*] *Report on Civil Aviation and Civil Government Air Operations for the Years 1927–1930*.

Coad, B.R., E. Johnson, and G.L. McNeil. *Dusting Cotton from Airplanes*. Washington, DC: United States Department of Agriculture, 1924.

Creighton, W. *Forestkeeping: A History of the Department of Lands and Forests in Nova Scotia, 1926–1960*. Halifax: Nova Scotia Department of Government Services, 1988.

Dominion of Canada, *Official Report of Debates*, House of Commons, 1925–1930.

Fernow, B.E. *Forest Conditions of Nova Scotia*. Ottawa: Commission of Conservation, 1912.

Gibson, A. *Report of the Dominion Entomologist, Arthur Gibson, F.R.S.C., F.E.S.A. for the Two Years 1919 and 1920*. Department of Agriculture. Ottawa: F.A. Acland, Printer to the King's Most Excellent Majesty, 1923.

Hawboldt, L.S. *The Spruce Budworm*. Halifax: Nova Scotia Department of Lands and Forests, 1955.

Hewitt, C.G. *The Control of Insect Pests in Canada*. Ottawa: Government Printing Bureau, 1912.

– *Report of the Dominion Entomologist, C.G. Hewitt, for the Year Ending March 31, 1910*. Department of Agriculture. Ottawa: Government Printing Bureau, 1911.

– *The Spruce Budworm and Larch Sawfly*. Kingston, ON: British Whig Publishing, 1911.

"The Hon. James Layton Ralston." PARLINFO. Accessed 26 September 2016. https://lop.parl.ca /sites/ParlInfo/default/en_CA/People/Profile?personId=8357.

Hopping, R. *The Control of Bark-Beetle Outbreaks in British Columbia*. Ottawa: Canada Department of Agriculture, Entomological Branch, Circular 15, 1921.

Information for Intending Settlers: Muskoka and Lake Nipissing Districts. Ottawa: Canada Department of Agriculture, 1880.

Johnson, R.S. *Forests of Nova Scotia: A History*. Halifax: Nova Scotia Department of Lands and Forests, 1986.

Lecky, C.S., and M.S. Murphy. *History of Tallulah Laboratory*. United States Department of Agriculture, Bureau of Entomology, June 1936.

Nova Scotia. *Report of the Department of Lands and Forests, 1925–1932*.

Ontario. *Report of the Minister of Lands and Forests, 1920–1932*.

Ontario GEOservices Centre and Ministry of Northern Development and Mines. *Rock ONtario*. Toronto: Queen's Printer for Ontario, 1994.

Prebble, M.L., ed. *Aerial Control of Forest Insects in Canada: A Review of Control Projects Employing Chemical and Biological Insecticides*. Ottawa: Department of the Environment, 1975.

Report of the Royal Commission on Pulpwood, Ottawa, July 1924. Ottawa: Printer to the King's Most Excellent Majesty, 1924.

Schmitt, D.M., D.G. Grimble, and J.L. Searcy. *Managing the Spruce Budworm in Eastern North America*. Washington, DC: United States Department of Agriculture, Forest Service, Cooperative State Research Service, 1984.

Swaine, J.M. *Canadian Bark-Beetles, Parts I–III: A Preliminary Classification, With an Account of the Habits and Means of Control*. Ottawa: J. de Labroquerie Tache, 1918.

– *Forest Entomology and Its Development in Canada, 1928*. Ottawa: F.A. Acland, Printer to the King's Most Excellent Majesty, 1928.

– *Forest Insect Conditions in British Columbia: A Preliminary Survey*. Ottawa: Government Printing Bureau, 1914.

– *The Hemlock Looper*. Ottawa: Minister of Agriculture, 1931.

– *The Spruce Budworm*. Ottawa: Minister of Agriculture, 1931.

Swaine, J.M., and F.C. Craighead. *Studies on the Spruce Budworm* (*Cacoecia fumiferana* Clem.). Ottawa: Canada Department of Agriculture, 1924.

United States Department of Agriculture. *Trees: The Yearbook of Agriculture*. Washington, DC: United States Government Printing Office, 1949.

– *Yearbook of Agriculture 1928*. Washington, DC: United States Government Printing Office, 1928.

United States Office of Experiment Stations. *Report of the Agricultural Experiment Stations, 1927*. Washington, DC: United States Department of Agriculture, 1927.

West, B. *The Firebirds: An Account of the First 50 Years of the Ontario Provincial Air Service*. Toronto: Ontario Ministry of Natural Resources, 1974.

Whitford, H.N., and R.D. Craig. *The Forest Resources of British Columbia*. Ottawa: Commission of
    Conservation, 1918.

NEWSPAPERS/PERIODICALS

Border Cities Star
British Columbia History
Canada Lumberman
Canada Lumberman and Woodworker
Canadian Woodlands Review
Christian Science Monitor
Flight Magazine
Forest Worker
Forestry Chronicle
Fredericton Gleaner
Globe (Toronto)
Halifax Herald
Journal of Economic Entomology
Journal of Forestry
Lewiston Daily Sun
London Times
Maclean's
Mail and Empire (Toronto)
Montreal Gazette
Montreal Star
Moody's Manual of Industrials
Pulp and Paper Magazine of Canada
Saturday Sunset
Scientific Agriculture
Sydney Record
The Bulletin (Canadian Airways Limited)
The Timberman
The (Vancouver) World
Vancouver Daily Province
Vancouver Star
Vancouver Sun
Wisconsin State Journal

Secondary Sources

Adam, G.M. *Muskoka Illustrated: With Descriptive Narrative of this Picturesque Region*. Toronto:
    Bryce, 1888.
"Aldo Leopold." The Aldo Leopold Foundation. Accessed 25 May 2019. https://www.aldoleopold
    .org/about/aldo-leopold.
Andrews, R. *Timber: Toil and Trouble in the Big Woods*. Seattle: Superior Publishing Company, 1968.
"Arthur Gibson." Find a Grave. Accessed 4 May 2018. https://www.findagrave.com/memorial
    /147596045/arthur-gibson.

Bastedo, W.A. *Materia Medica: Pharmacology, Therapeutics and Prescription Writing for Students and Practitioners*. Philadelphia: W.B. Saunders, 1918.

Baxter, J.S., and E.J. Hall, eds. "The Muskoka Lakes Highlands of Ontario, Canada." In *America: Her Grandeur and Her Beauty*. Chicago: Union Book, 1904.

Benedickson, J. *The Culture of Flushing: A Social and Legal History of Sewage*. Vancouver: UBC Press, 2007.

Benjamin, D.M., and D.W. Renlund. *Insecticide Use in Wisconsin Natural Forests and Plantations, 1969–1976*. Forestry Research Notes 198. Madison, WI: University of Wisconsin, 1976.

Bliss, M. *Northern Enterprise: Five Centuries of Canadian Business*. Toronto: McClelland & Stewart, 1987.

Boyer, G.W. *Early Days in Muskoka: A Story About the Settlement of Communities in the Free Grant Lands and of Pioneer Life in Muskoka*. Bracebridge, ON: Herald-Gazette Press, 1970.

Brown, R.C., and R. Cook. *Canada: 1896–1921, A Nation Transformed*. Toronto: McClelland & Stewart, 1974.

Buhs, J.B. *The Fire Ant Wars: Nature, Science, and Public Policy in Twentieth Century America*. Chicago: University of Chicago Press, 2004.

Cahill, B. *The Thousandth Man: A Biography of James McGregor Stewart*. Toronto: University of Toronto Press, 2000.

Campbell, C. *Shaped by the West Wind: Nature and History in Georgian Bay*. Vancouver: UBC Press, 2004.

Carson, R. *The Sea Around Us*. New York: Oxford University Press, 1951.

– *Silent Spring*. Boston: Houghton Mifflin, 1962.

Castonguay, S. "The Emergence of Research Specialties in Economic Entomology in Canadian Government Laboratories after World War II." *Historical Studies in Physical and Biological Sciences* 32, no. 1 (2001): 19–40.

– "Naturalizing Federalism: Insect Outbreaks and the Centralization of Entomological Research in Canada, 1884–1914." *Canadian Historical Review* 85, no. 1 (March 2004): 1–34.

– *Protection des cultures, construction de la nature: Agriculture, foresterie et entomologie au Canada, 1884–1959*. Saint-Laurent, QC: Diffusion Dimedia, 2004.

Cook, G.M. "'Spray, Spray, Spray!': Insecticides and the Making of Applied Entomology in Canada, 1871–1914." *Scientia Canadensis* 22–3 (1998–9): 7–50.

Creighton, D. *The Commercial Empire of the St. Lawrence, 1760–1850*. Toronto: Ryerson Press, 1937.

Cronon, W. "The Trouble with Wilderness: Or, Getting Back to the Wrong Nature." *Environmental History* 1, no. 1 (January 1996): 7–28.

Cruikshank, K. *Close Ties: Railways, Government and the Board of Railway Commissioners, 1851–1933*. Montreal: McGill-Queen's University Press, 1991.

Cruikshank, K., and N. Bouchier. *The People and the Bay: A Social and Environmental History of Hamilton Harbour*. Vancouver: UBC Press, 2016.

Deschouwer, K., and M.T. Jans, eds. *Politics Beyond the State: Actors and Policies in Complex Institutional Settings*. Brussels: VUBPress, 2007.

de Vecchi, M.G. "Science and Government in Nineteenth-Century Canada." PhD diss., University of Toronto, 1978.

Douglas, W.A.B. *The Creation of the National Air Force: The Official History of the Royal Canadian Air Force, Vol. II*. Toronto: University of Toronto Press, 1986.

Downs, E.W., and G.F. Lemmer. "Origins of Aerial Crop Dusting." *Agricultural History* 39, no. 3 (July 1965): 123–35.

Drummond, I.M. *Progress without Planning: The Economic History of Ontario from Confederation to the Second World War*. Toronto: University of Toronto Press, 1987.

Drushka, K. *HR: A Biography of H.R. MacMillan*. Madeira Park, BC: Harbour Publishing, 1995.

Dunlap, T.R. *DDT: Scientists, Citizens and Public Policy*. Princeton, NJ: Princeton University Press, 1981.

Entomological Society of British Columbia. *Proceedings for 1941*.

Evans, C.L. *The War on Weeds in the Prairie West: An Environmental History*. Calgary: University of Calgary Press, 2002.

Favareau, D. "Understanding Natural Constructivism." *Semiotica* 172 (November 2008): 489–528.

Forster, B. *A Conjunction of Interests: Business, Politics, and Tariffs, 1825–1879*. Toronto: University of Toronto Press, 1986.

Fracker, S.B., and A.A. Granovsky. "Airplane Dusting to Control the Hemlock Spanworm." *Journal of Forestry* 26, no. 1 (1928): 12–33.

– "The Control of the Hemlock Spanworm by Airplane Dusting." *Journal of Economic Entomology* 20 (1927): 287–98.

Gandy, M. *Concrete and Clay: Reworking Nature in New York City*. Cambridge, MA: MIT Press, 2002.

Gibson, A. "The Canadian Entomological Service: Fifty Years of Retrospect, 1887 to 1937." In *Internationaler Kongreb fur Entomologie VII, Berlin, 15–20 August 1938*. Weimar: G. Uschmann, 1939, 1429–79.

Gillis, P.R., and T.R. Roach. *Lost Initiatives: Canada's Forest Industries, Forest Policy and Forest Conservation*. New York: Greenwood Press, 1986.

Girard, M.F. *L'écologisme retrouvé: Essor et déclin de la Commission to la conservation du Canada, 1909–1921*. Ottawa: Presses de l'université d'Ottawa, 1994.

Graham, R. *The King Byng Affair, 1926*. Toronto: Copp Clark, 1967.

Gray, S. "The Government Timber Business: Forest Policy and Administration in British Columbia, 1912–1928." *BC Studies* 81 (Spring 1989): 24–49.

Greenhous, B., and H. Halliday. *Canada's Air Forces, 1914–1999*. Montreal: Editions Art Global and the Department of National Defence, 1999.

Hays, S.P. *Beauty, Health and Permanence: Environmental Politics in the United States, 1955–1985*. New York: Cambridge University Press, 1987.

– *Conservation and the Gospel of Efficiency: The Progressive Conservation Movement, 1890–1920*. Cambridge, MA: Harvard University Press, 1959.

"Hemlock Looper." National Resources Canada. Accessed 18 November 2016. https://tidcf.nrcan .gc.ca/en/insects/factsheet/8846.

Hidy, R.W., F.E. Hill, and A. Nevins. *Timber and Men: The Weyerhaeuser Story*. New York: Macmillan, 1963.

Hodgins, B.W., and J. Benedickson. *The Temagami Experience: Recreation, Resources, and Aboriginal Rights in the Northern Ontario Wilderness*. Toronto: University of Toronto Press, 1989.

Hosie, R.C. *Native Trees of Canada*. 8th ed. Markham, ON: Fitzhenry & Whiteside, 1990.

"Huff-Daland-Duster." Delta Flight Museum. Accessed 9 December 2011. https://www.delta museum.org/exhibits/delta-history/aircraft-by-type/crop-duster/Huff-Daland-Duster.

Humphreys, P., and S.G. Wong. *The History of Wigwam Inn*. Vancouver: Perfect Printers, 1982.

Jacques Cattell Press, ed. *American Men and Women of Science, Vol. 5, P–Sr*. 12th ed. New York and London: Jacques Cattell Press and R.R. Bowker Company, 1972.

"James Malcolm Swaine, 1878–1955." *Canadian Entomologist* 87, no. 10 (October 1955): 460.

Jansen, S. "Chemical-Warfare Techniques for Insect Control: Insect 'Pests' in Germany Before and After World War I." *Endeavour* 24, no. 1 (March 2000): 28–33.

Jasen, P. *Wild Things: Nature, Culture and Tourism in Ontario, 1790–1914*. Toronto: University of Toronto Press, 1995.

Kheraj, S. *Inventing Stanley Park: An Environmental History*. Vancouver: UBC Press, 2013.

– "Restoring Nature: Ecology, Memory, and the Storm History of Vancouver's Stanley Park." *Canadian Historical Review* 88, no. 4 (December 2007): 577–612.

Krech, S. "Fire." In *Canadian Environmental History: Essential Readings*, edited by D.F. Duke. Toronto: Canadian Scholars' Press, 2006, 115–35.

Kuhlberg, M. "'An Eden that is practically uninhabited by humans': Manipulating Wilderness in Managing Vancouver's Drinking Water, 1880–1930." *Urban History Review* 45, no. 1 (Fall 2016): 18–35.

– *In the Power of the Government: The Rise and Fall of Newsprint in Ontario, 1894–1932*. Toronto: University of Toronto Press, 2015.

– *One Hundred Rings and Counting: Forestry Education and Forestry in Toronto and Canada, 1907–2007*. Toronto: University of Toronto Press, 2009.

– "'We are the pioneers in this business': Spanish River's Forestry Initiatives After the Great War." *Ontario History* 93, no. 2 (Fall 2001): 151–78.

Leane, J.J. *The Oxford Story: A History of the Oxford Paper Company, 1847–1958*. Rumford, ME: Oxford Paper Company, 1958.

Legge, R.M. *Sawdust and Sea Breezes: A History of Liscomb Mills, Guysborough County*. Antigonish, NS: Casket Printing and Publishing, 2005.

Leopold, A. *A Sand County Almanac*. New York: Oxford University Press, 1966.

Loo, T. *States of Nature: Conserving Canada's Wildlife in the Twentieth Century*. Vancouver: UBC Press, 2006.

Lower, A.R.M. *Great Britain's Woodyard: British America and the Timber Trade, 1763–1867*. Montreal: McGill-Queen's University Press, 1973.

– *The North American Assault on the Canadian Forest: A History of the Lumber Trade Between Canada and the United States*. New York: Greenwood Press, 1938.

– *Settlement and the Forest Frontier in Eastern Canada*. Toronto: Macmillan Company of Canada, 1936.

Lundell, L. *Old Muskoka: Century Cottages & Summer Estates*. Toronto: Boston Mills Press, 2003.

Macdonald, D. *Business and Environmental Politics in Canada*. Toronto: University of Toronto Press, 2007.

MacEachern, A. *The Miramichi Fire: A History*. Montreal: McGill-Queen's University Press, 2020.

– *Natural Selections: National Parks in Atlantic Canada, 1935–1970*. Montreal: McGill-Queen's University Press, 2001.

Macfarlane, D. "'A Completely Man-Made and Artificial Cataract': The Transnational Manipulation of Niagara Falls." *Environmental History* 18, no. 4 (October 2013): 759–84.

Mart, M. *A Love Story: America's Enduring Embrace of Dangerous Chemicals*. Lawrence: University of Kansas Press, 2016.

McLaughlin, M.J. "Green Shoots: Aerial Insecticide Spraying and the Growth of Environmental Consciousness in New Brunswick, 1952–1973." *Acadiensis* 40, no. 1 (Winter/Spring 2011): 3–23.

McMurray, T. *The Free Grant Lands of Canada: From Practical Experience to Bush Farming in the Free Grant Districts of Muskoka and Parry Sound*. Bracebridge, ON: Northern Advocate, 1871.

Moody's Manual of Industrials, 1920s. New York: Moody's, 1920s.

Muskoka: Land of Health and Pleasure. Toronto[?], 1896[?].

Nash, L. "The Changing Experience of Nature: Historical Encounters with a Northwest River." *Journal of American History* 86, no. 4 (March 2000): 1600–29.

Nash, R. *Wilderness and the American Mind*. New Haven, CT: Yale University Press, 1967.

Natural Resources Canada. "Spruce Budworm (Factsheet)." Government of Canada. Accessed 18 November 2016. http://www.nrcan.gc.ca/forests/fire-insects-disturbances/top-insects /13403.

Neillie, C.R., and J.S. Houser. "Fighting Insects with Airplanes: An Account of the Successful Use of the Flying-Machine in Dusting Tall Trees Infested with Leaf-Eating Caterpillars." *National Geographic Magazine* 41, no. 3 (March 1922): 333–8.

Parenteau, B., and L.A. Sandberg. "Conservation and the Gospel of Economic Nationalism: The Canadian Pulpwood Question in Nova Scotia and New Brunswick, 1918–1925." *Environmental History Review* 19, no. 2 (Summer 1995): 31–58.

Parr, J. *Sensing Changes: Technologies, Environments, and the Everyday, 1953–2003*. Vancouver: UBC Press, 2010.

Picturesque Canada: The Northern Lakes Guide. Toronto: Hunter, Rose & Co., 1879.

Pigott, P. *Taming the Skies: A Celebration of Canadian Flight*. Toronto: Dundurn Press, 2008.

Post, G.B. "Boll Weevil Control by Airplane." *Georgia State College of Agriculture Extension Division Bulletin 301* 13, no. 4 (November 1924): 1–22.

Proceedings of the 22nd Annual Meeting of the Entomological Society of Alberta, October 3–5 1974, Volume 22, February 1975.

Pyne, S.J. *Awful Splendour: A Fire History of Canada*. Vancouver: UBC Press, 2007.

– *Fire in America: A Cultural History of Wildland and Rural Fire*. Seattle: University of Washington Press, 2004.

Rajala, R.A. "The Vernon Laboratory and Federal Entomology in British Columbia." *Journal of Entomological Society of British Columbia* 98 (December 2001): 177–88.

Reeves, A. "In Stanley Park, You Say? Killing Caterpillars on a Sunday." *British Columbia History* 39, no. 1 (2006): 2–3.

Richmond, H.A. "A History of Forest Entomology in British Columbia, 1920–1984." *British Columbia Forest History Newsletter*, no. 9 (November 1984): 4–7.

Riegert, P.W. "Charles Hewitt Gibson." In *Dictionary of Canadian Biography Volume XIV, 1911–1920*, edited by R. Cook and J. Hamelin. Toronto: University of Toronto Press, 1998.

– *From Arsenic to DDT: A History of Entomology in Western Canada*. Toronto: University of Toronto Press, 1980.

– "James Fletcher." In *Dictionary of Canadian Biography Volume XIII, 1901–1910*, edited by R. Cook and J. Hamelin. Toronto: University of Toronto Press, 1994.

Roach, T.R. "Stewards of the People's Wealth: The Founding of British Columbia's Forest Branch." *Journal of Forest History* 28, no. 1 (January 1984): 14–23.

Roach, T.R., and R. Judd. "A Man for All Seasons: Frank John Dixie Barnjum, Conservationist, Pulpwood Embargoist and Speculator!" *Acadiensis* 20, no. 2 (Spring 1999): 129–44.

Rogers, J. *Guide Book and Atlas of Muskoka and Parry Sound Districts*. Toronto: H.R. Page, 1879.

– *Muskoka Lakes Blue Book: Directory and Chart, 1918*. Port Sandfield, ON: John Rogers, 1918.

Russell, E. "Speaking of Annihilation: Mobilizing for War against Humans and Insect Enemies, 1914–1945." *Journal of American History* 82, no. 4 (1996): 1505–29.

– "The Strange Career of DDT: Experts, Federal Capacity, and Environmentalism in World War II." *Technology and Culture* 40, no. 4 (October 1999): 770–96.

– *War and Nature: Fighting Humans and Insects with Chemicals from World War II to Silent Spring*. New York: Cambridge University Press, 2001.

Sandberg, L.A. "Forest Policy in Nova Scotia: The Big Lease, Cape Breton Island, 1899–1960." *Acadiensis* 20, no. 2 (Spring 1991): 105–28.

Sandberg, L.A., and P. Clancy. "Politics, Science and the Spruce Budworm in New Brunswick and Nova Scotia." *Journal of Canadian Studies* 37, no. 2 (Summer 2002): 164–91.

Schierbeck, O. *Treatise on the Spruce Bud Worm, Bark Beetle and Borer*. Montreal: F.J.D. Barnjum, 1922.

Schneider, D. *Hybrid Nature: Sewage Treatment and the Contradictions of the Industrial Ecosystem*. Cambridge, MA: MIT Press, 2011.

Spirn, A.W. "Constructing Nature: The Legacy of Frederick Law Olmsted." In *Uncommon Ground: Rethinking the Human Place in Nature*, edited by W. Cronon. New York: W.W. Norton, 1996, 91–113.

Bibliography

Thistle, J. "Accommodating Cattle: British Columbia's 'Wars' with Grasshoppers and 'Wild Horses.'" *BC Studies* 160 (Winter 2008–9): 67–91.

Traill, C.P. *The Backwoods of Canada: Being Letters from the Wife of an Emigrant Officer, Illustrative of the Domestic Economy of British America*. London: Charles Knight & Co., 1846.

Traves, T. *The State and Enterprise: Canadian Manufacturers and the Federal Government, 1917–1931*. Toronto: University of Toronto Press, 1979.

Turner, K.B. "A History of Investigations into and Control of Pests of Forest Trees in Ontario." First Draft. Ontario Forest Protection Branch, September 1965.

Washington Forest Protection Association. *A Chronology of the First 100 Years of the Washington Forest Protection Association, 1908–2008*.

White, R. *The Organic Machine: The Remaking of the Columbia River*. New York: Hill and Wang, 1995.

Whorton, J.C. *Before Silent Spring: Pesticides and Public Health in Pre-DDT America*. Princeton, NJ: Princeton University Press, 1974.

– "Insecticide Spray Residues and Public Health: 1865–1938." *Bulletin of the History of Medicine* 45, no. 3 (May 1971): 219–41.

Wickman, B.E., T.R. Torgensen, and M.M. Furniss. "Photographic Images and History of Forest Investigations on the Pacific Slope, ca. 1910–1953, Part 2, Oregon and Washington." *American Entomology* 48, no. 3 (2002): 178–85.

Wiegman, C. *Trees to News: A Chronicle of the Ontario Paper Company's Origins and Development*. Toronto: McClelland & Stewart, 1953.

Wilson, E. "The Use of Aircraft in Forestry and Logging." *Aeroplane* 19, no. 18 (November 1920): 730–6.

– "Through Canadian Wilds: Three Sketches of Early Forestry in Quebec." *Forest & Conservation History* 11, no. 4 (January 1968): 16–25.

World Forestry Center. "Charles S. Cowan: Biographical Portrait (1887–1969)." *Forest History Today* (Spring 2012): 28–30.

Zeller, S. "Darwin Meets the Engineers: Scientizing the Forest at McGill University, 1890–1910." *Environmental History* 6, no. 3 (July 2001): 428–50.

– *Inventing Canada: Early Victorian Science and the Idea of a Transcontinental Nation*. Toronto: University of Toronto Press, 1987.

– *Land of Promise, Promised Land: The Culture of Victorian Science in Canada*. Ottawa: Canadian Historical Association, 1996.

Index